NUMERICAL METHODS

FOR ENGINEERS AND COMPUTER SCIENTISTS

NUMERICAL METHODS

FOR ENGINEERS AND COMPUTER SCIENTISTS

PAUL F. HULTQUIST

University of Colorado at Denver

The Benjamin/Cummings Publishing Company, Inc.

Menlo Park, California • Reading, Massachusetts
Don Mills, Ontario • Wokingham, U.K. • Amsterdam • Sydney
Singapore • Tokyo • Madrid • Bogota • Santiago • San Juan

Sponsoring Editor: Sally Elliott
Production Editor: Wendy Calmenson, The Book Company
Copyeditor: Linda Thompson
Cover Designer: Wendy Calmenson, The Book Company
Cover Photo: N. Foster, © The Image Bank
Technical Artists: Reese and Deborah Thornton, Folium
Composition: Graphic Typesetting Service

The basic text of this book was designed using the Modular Design System, as developed by Wendy Earl and Design Office Bruce Kortebein.

Library of Congress Cataloging-in-Publication Data

Hultquist, Paul F.
  Numerical methods for engineers and computer scientists / Paul F. Hultquist.
    p.   cm.
  Includes index.
  ISBN 0-8053-4652-X
  1. Numerical analysis—Data processing. 2. Engineering mathematics—Data processing. I. Title.
QA297.H85 1988
519.4—dc19

ISBN: 0-8053-4652-X
ABCDEFGHIJ-DO-8987

The Benjamin/Cummings Publishing Company, Inc.
2727 Sand Hill Road
Menlo Park, California 94025

Preface

Purpose

As science and engineering become more and more sophisticated, there is an increasing need for practitioners to develop skills in using the computer. The necessary skills go beyond the ability to program and include the ability to formulate problems correctly and to solve problems requiring substantial amounts of computation. Because there is now good numerical software available, it is no longer efficient for persons needing to use the computer to write their own numerical procedures. Rather, they should be knowledgeable about the commonly used numerical algorithms and why, where, and when these algorithms succeed or fail. The underlying philosophy of this book is to try to create understanding about the algorithms and not to belabor the fine points of the construction of them. In a true sense it is a book on numerical methods rather than one on numerical analysis. Nonetheless, it can be used in a mathematics department to introduce mathematics majors to the computational mathematics field with the notion that they may be motivated to study the subject later in a more rigorous way. The target audiences are junior and senior students in the sciences and in engineering who need to learn about numerical methods in a one-semester course and professionals who need to upgrade their computer problem-solving skills.

Features

The official language chosen is PASCAL. It is widely taught and it is a well-structured and strongly typed language that encourages good programming. There are PASCAL procedures in the book for all the major topics, and these are available on floppy disks for individual student use on personal computers. For those who prefer FORTRAN we encourage the use of good program libraries, notably IMSL. An IMSL subroutine is suggested for each major topic, and examples are shown for the use of some of these subroutines.

Problem sets are given in each chapter. One of the features of the book is the inclusion of *project problems*. These are real-life kinds of problems, many of which are adapted from work that the author has performed in industry or as a consultant. The author generally has required the solution of five to seven of these as a major part of the one-semester course. The materials required to be handed in included a program listing, printed (or graphical) output, and a brief report. The instructor should pay careful attention to such matters as physical units, the correctness of the results, the accuracy, style, spelling, and grammar of the report, and the organization of the output. If the course in which the book is to be used is a junior or senior engineering course, then it should have a design emphasis with respect to the software. The student should be encouraged to do a professional job on all of the project reports.

Prerequisites

Prerequisites for the course include mathematics through analytic geometry and calculus, an introduction to ordinary differential equations, and an introduction to linear algebra. The last two courses are combined into a single course in many schools, which is a minimum satisfactory preparation. The standard physics and chemistry courses, of course, are necessary background for more advanced work in some field of science or engineering; lack of advanced background in mechanics, electricity, or heat and thermodynamics may make it difficult to solve some of the project problems without help from an instructor.

Acknowledgments

Finally, I wish to thank the many persons who have helped me in the creation of this book. Paul Saylor of the University of Illinois, David S. Scott of the University of Texas and now of Intel, and Stanley L. Steinberg of the University of New Mexico read the first draft of the manuscript and were most helpful in suggesting substantial improvements. James Hummel of the University of Maryland, Donald Grace of Oklahoma State University, and Thomas Foley of Arizona State University read the second draft and were very helpful in suggesting changes and improvements. Alan Apt, who encouraged me to submit a book proposal, Craig Bartholomew, who patiently guided me through two revisions of the manuscript, and Sally Elliott of Benjamin/Cummings deserve special thanks, as well as Wendy Calmenson of The Book Company, who handled the book production.

Craig Schons, a former student, helped significantly by writing several of the programs as an independent study project. Colleagues in my department deserve thanks for the encouragement they gave me, in particular Bill Murray, who was chairman at the time I started on the project. My family, Juanita, Fred, and Ann, encouraged and supported me through the long days and nights of writing and rewriting. Also, I shall be forever grateful to a host of unnamed friends without whose support I would not have been able to finish this book.

Lakewood, Colorado
August 1, 1987

Contents

NUMERICAL METHODS

FOR ENGINEERS AND COMPUTER SCIENTISTS

1

Numbers and Their Representations

Because numbers are the lifeblood of scientific and engineering computing, it is necessary for us to understand how they are represented in the computer and how they are treated in numerical computations. Central to that understanding is the whole matter of number systems.

1.1 Number Bases

In the system of number representation with which we are familiar (called a *positional system*), each digit's value is determined by its position with respect to the decimal point. For example, the number 675 means 6 times 100 plus 7 times 10 plus 5, so that the digits carry different values depending upon their position within the number. Contrast this with the Roman numeral system, where a symbol has a value more or less independent of its position. However, in the case where a smaller numeral precedes a larger one, the smaller numeral has a negative value:

$$IV = -1 + 5 = 4$$
$$VI = 5 + 1 = 6$$

It is extremely difficult to do arithmetic in the Roman numeral system, and one of its few uses in recent times is in representing dates, such as on the cornerstones of public buildings.

The Arabic system of numerals (symbols borrowed from the Sanskrit) with positional representation leads to relatively easy rules of arithmetic. *Algorithms* (prescriptions for carrying out some task that can be described in systematic terms) for doing arithmetic in the decimal system have been known for many centuries.

The idea of other number bases than 10, once thought mainly to be a curiosity, has become an important one in the last generation. John von Neumann suggested during World War II that the binary (base 2) number system was the natural one for computers, because it allowed the use of enormously simplified electronic circuitry of high reliability.* Binary systems are now almost universally used in computing, although as we shall see later, groups of three or four binary digits can be grouped to form octal (base 8) or hexadecimal (base 16) digits for convenience.

Every positional number system has the same features:

1. A base b, such as 2 (binary), 8 (octal), 10 (decimal), 16 (hexadecimal).
2. A set of b symbols, such as $\{0, 1\}$, $\{0, 1, 2, 3, 4, 5, 6, 7\}$, $\{0, 1, 2, 3, 4, 5, 6, 7, 8, 9\}$, or $\{0, 1, \ldots , 9, A, B, C, D, E, F\}$. (Notice in the hexadecimal system, $A = 10$, $B = 11$, $C = 12$, and so on.)
3. Positional representation:

$$a_n a_{n-1} \cdots a_1 a_0 . a_{-1} a_{-2} \cdots$$

where the a's are members of the set of symbols and where a_k carries as its value $a_k b^k$.

Thus $225.3 = 2 \times 10^2 + 2 \times 10^1 + 5 \times 10^0 + 3 \times 10^{-1}$. Likewise, the binary number

$$101.11 = 1 \times 2^2 + 0 \times 2^1 + 1 \times 2^0 + 1 \times 2^{-1} + 1 \times 2^{-2}$$

which is 5.75 in decimal.

In order to distinguish 101.11 (binary) from 101.11 (decimal) we often write

$$101.11_2 = 5.75_{10}$$

or

$$(101.11)_2 = (5.75)_{10}$$

1.1.1 Conversion Among Bases

The previous discussion gives a hint of how conversions between number bases can be done. Basically, there are four kinds of conversions needed. A general conversion of a number with an integer part and a fraction requires the use of

---

*There is evidence that John Atanasoff independently came to the same conclusion before World War II. Because he was not successful in creating a complete working computer that attracted wide attention, his name did not become attached to the idea.

two of the four methods. The four methods depend upon whether the conversion is made from the decimal system to some other system, or vice versa, and on whether the number being converted is a fraction or an integer. These methods will be illustrated using octal and decimal systems, but the principles are the same for conversions among any two number systems.

Converting Octal Integers to Decimal Integers   This method is easily understood by means of an example.

$$257_8 = 2 \times 8^2 + 5 \times 8 + 7 = 128 + 40 + 7 = 175_{10}$$

The conversion is more efficiently done in *nested* form:

$$(2 \times 8 + 5) \times 8 + 7 = 21 \times 8 + 7 = 168 + 7 = 175_{10}$$

Converting Octal Fractions to Decimal Fractions

$$(0.257)_8 = 2 \times 8^{-1} + 5 \times 8^{-2} + 7 \times 8^{-3} = 0.341796875_{10}$$

This is also done more efficiently in nested form:

$$\left(\left(\tfrac{7}{8} + 5\right)\big/8 + 2\right)\big/8 = (5.875/8 + 2)/8 = 2.734375/8 = 0.341796875_{10}$$

Converting Decimal Integers to Octal Integers   The algorithm is easy to understand if we recognize that, for example,

$$375_{10} = a_0 + a_1 \cdot 8 + a_2 \cdot 8^2 + \cdots$$

We now divide both sides by 8:

$$46 + \tfrac{7}{8} = \frac{a_0}{8} + a_1 + a_2 \cdot 8 + \cdots$$

The integer parts must equal the integer parts and the fractions must equal the fractions, which allows us to conclude that $a_0 = 7$ and that

$$46_{10} = a_1 + a_2 \cdot 8 + a_3 \cdot 8^2 + \cdots$$

Division by 8 again yields

$$5 + \tfrac{6}{8} = \frac{a_1}{8} + a_2 + a_3 \cdot 8 + \cdots$$

Hence $a_1 = 6$ and

$$5 = a_2 + a_3 \cdot 8 + a_4 \cdot 8^2 + \cdots$$

Again, dividing by 8, we have

$$0 + \frac{5}{8} = \frac{a_2}{8} + a_3 + a_4 \cdot 8 + \cdots$$

which allows us to write $a_2 = 5$ and

$$0 = a_3 + a_4 \cdot 8 + \cdots$$

This can be true if and only if $a_3 = a_4 = \cdots = 0$. Thus

$$375_{10} = (a_2a_1a_0)_8 = 567_8$$

In a shortcut form, this method is as follows:

$$
\begin{array}{r}
46 \\
\hline
8\ )375 \\
32 \\
\hline
55 \\
48 \\
\hline
7 = a_0
\end{array}
\qquad
\begin{array}{r}
5 = a_2 \\
\hline
8\ )46 \\
40 \\
\hline
6 = a_1
\end{array}
\qquad (\text{because } 5 < 8)
$$

Converting Decimal Fractions to Octal Fractions   The method here is analogous to that of the last method, except that we use multiplication. For example, let us convert decimal 0.35 to octal:

$$0.35_{10} = a_{-1} \cdot 8^{-1} + a_{-2} \cdot 8^{-2} + a_{-3} \cdot 8^{-3} + \cdots$$

Multiply by 8:

$$2.80 = a_{-1} + a_{-2} \cdot 8^{-1} + a_{-3} \cdot 8^{-2} + \cdots$$

Because the whole number part on the left equals the whole number a_{-1}, we can subtract $a_{-1} = 2$ from both sides and multiply again:

$$6.40 = a_{-2} + a_{-3} \cdot 8^{-1} + a_{-4} \cdot 8^{-2} + \cdots$$

Now $a_{-2} = 6$. Subtract this from both sides and continue:

$$3.20 = a_{-3} + a_{-4} \cdot 8^{-1} + a_{-5} \cdot 8^{-2} + \cdots$$

Thus $a_{-3} = 3$. If we continue, we find that $a_{-4} = 1$, $a_{-5} = 4$, and so on, so that

$$0.35_{10} = 0.2631463146 \cdots_8$$

This process does not terminate as in earlier processes. Here we have an example of one of the problems we face in numerical computation: numbers that are represented by terminating decimals in one number base system may not be exactly represented in another base system because we have only finite-length registers in the computer.

1.1.2   Decimal, Binary, Octal, and Hexadecimal

Because humans have four fingers and a thumb on each hand, we are stuck with the decimal system, at least for now. Those of us who deal with computing machines are also stuck with at least three other common bases: binary, octal, and hexadecimal. That is not quite as bad as it seems because binary, octal, and hexadecimal are in a sense all interchangeable. Suppose we write a large binary number such as $10101110101101_2 = 11181_{10}$ and group the digits by 3s (adding 1 zero to the leftmost group), starting from the right end:

010__101__110__101__101

Now write each triplet in decimal (which has the same set of digits as octal for 0 through 7):

2__5__6__5__5

This is simply the octal number 25655.

If we group the binary digits in 10101110101101 by fours (adding 2 zeros to the leftmost group) and replace each group by hexadecimal digits, we have

$$0010\_\_1011\_\_1010\_\_1101 = 2BAD_{16}$$

where $A = 10 = 1010_2$, $B = 11 = 1011_2$, and $D = 13 = 1101_2$.

If we write a binary integer in general as

$$\sum_{k=0}^{n} a_k 2^k = a_n \cdot 2^n + a_{n-1} \cdot 2^{n-1} + \cdots + a_1 \cdot 2 + a_0$$

we see that successive 3-bit (binary integer) groups (counted from the right-hand end of a number) are of the form

$$a_{3m+2} \cdot 2^{3m+2} + a_{3m+1} \cdot 2^{3m+1} + a_{3m} \cdot 2^{3m}$$

where $m = 0$ for the rightmost 3 bits, $m = 1$ for the 3 bits to the left of those, and so on. We can factor $2^{3m} = 8^m$ out of each of the terms to obtain

$$(a_{3m+2} \cdot 2^2 + a_{3m+1} \cdot 2 + a_{3m}) \cdot 8^m$$

The quantity in parentheses is simply an octal digit, which is multiplied by the appropriate power of 8. Thus

$$\sum_{k=0}^{n} a_k 2^k = \sum_{j=0}^{\lceil n/3 \rceil} b_j 8^j$$

where $\lceil n/3 \rceil$ means the least integer equal to or greater than $n/3$ and where the b_j are the octal digits obtained by grouping the binary digits by 3s.

A similar kind of proof can be done for the hexadecimal case using groups of 4 bits instead of groups of 3 bits.

All computer systems provide input-output routines that convert internal binary to decimal for printing or for cathode-ray tube (CRT) monitor display. However, for many system functions it is easier to print or display octal or hex digits rather than decimal. The conversions from binary to octal are simple. For certain purposes the individual bits may be significant and may represent states (on/off) of equipment or system functions. The use of byte-oriented machines (1 byte = 8 bits) has tended to increase the importance of hexadecimal over octal. A byte holds exactly 2 hex digits, whereas octal digits will fit only those systems having word lengths that are multiples of 3 bits.

1.2   Number Representation

Computers are capable of representing numbers in several different ways. Integers are stored in one particular form, but numbers possessing fractional parts are stored in a different form. Although the actual form of storage is binary, for clarity we will use decimal in some of the discussions. The principles in each case are the same.

1.2.1   Integer Representation

The previous material dealt mostly with positive numbers. However, we need to be able to deal with negative-number representations in the computer as well. In the case of integers, there are three different methods of representing negative numbers: sign-magnitude, 1's complement, and 2's complement. The purpose of using complemented (negative) numbers is that they can be added in a normal adder, which eliminates the need for a subtractor. Complementation is simple to do in the hardware, and the ability to complement is necessary for some of the computer's other logic operations. Sign-magnitude computers also use complementation but at the actual time of addition or subtraction.

In most machines the most-significant bit of the computer word is reserved for the sign. Consequently, in the simplest case of sign-magnitude form, the decimal number -11 is in 1-byte form:

$$1\ \ 0\ \ 0\ \ 0\ \ 1\ \ 0\ \ 1\ \ 1$$
$$\uparrow\text{——— Sign bit}$$

In 1's complement, all the bits are simply complemented (i.e., each 1 bit is converted to 0, and each 0 bit is converted to 1), or it can be considered that the magnitude of the number is subtracted from a word filled with 1 bits:

$$1\,1\,1\,1\,1\,1\,1\,1 - 1\,0\,1\,1 = 1\,1\,1\,1\,0\,1\,0\,0$$

Note that the most-significant bit is a 1 to designate a negative number. Unused bit positions at the left end are filled with 1s in what is sometimes called *sign extension*. In the case of 2's complementation, the number is subtracted from all 1s plus 1:

$$100000000 = 11111111 + 1$$
$$\underline{ -1011}$$
$$11110101 = -11_{10}$$

In this case it is clear that the 2's complement is 1 plus the 1's complement. This is very little harder to implement in hardware than the 1's complement.

We will show a few examples of what happens in the hardware when complemented numbers are used.

In 2's complement,

$$11110101 = -11$$
$$\underline{00001011} = +11$$
$$1\!\mid\!00000000 = 0$$

We throw away the 1 that overflows the adder; it was simply the original 100000000 that was added to the negative number during the process of complementation. Other examples are as follows:

$$11110101 = -11$$
$$\underline{00001100} = +12$$
$$1\!\mid\!00000001 = +1$$

$$11110101 = -11$$
$$\underline{00001010} = +10$$
$$11111111 = -1 \quad \text{(Stays complemented)}$$

The preceding examples are in 2's complement. However, in 1's complement,

$$11110100 = -11$$
$$\underline{00001011} = +11$$
$$11111111 = 0 \quad (!)$$

This is the so-called negative zero, which is the 1's complement of *positive zero*. This form of arithmetic pays the penalty for its simplicity in forming complements by having two varieties of representation for zero.

Forming a positive sum involving a 1's complemented negative number results in a simple but puzzling rule:

$$11110100 = -11$$
$$\underline{00001100} = +12$$
$$\llbracket 1\!\mid\!00000000$$
$$\xrightarrow{\hspace{2cm}} 1 \quad \text{(End-around carry)}$$
$$\underline{}$$
$$00000001 = 1$$

The *end-around carry* is justified by the set of arithmetic operations shown next:

$$-1011 + 1100 = (11111111 - 1011) + 1100 - 11111111$$
$$= 11110100 + 1100 - 11111111 + (1 - 1)$$
$$= (11110100 + 1100) - (11111111 + 1) + 1$$
$$= 100000000 - 100000000 + 1 = 1$$

Note that complementation involves adding 11111111 to the negative number, and this must be subtracted out again to have a true result for a positive sum. The subtraction is done using 11111111 in the form of $100000000 - 1$.

We now need to look at how numbers with fractional parts are stored in the computer. Some of the same ideas, such as the use of complementation, still apply.

1.2.2   Floating-point Representation

The representation of a number with a fractional part is an adaptation of scientific notation. In scientific notation, we represent 245.6 as 2.456×10^2 and -0.00379 as -3.79×10^{-3}. The number is written as $P \times 10^Q$, where $1 \leqslant |P| < 10$ and Q is the appropriate integer such that

$$P \times 10^Q = N$$

where N is the number to be represented. Inside the computer we may wish to reorganize the way a number is stored, and we do this by declaring a variable to be of type REAL, which generally means that a machine representation known as *floating point* is used. Floating-point representations vary from machine to machine, but the following discussion, based on decimal, illustrates the principles involved.

The computer word is divided into sections called the *sign*, the *exponent*, and the *fraction* (sometimes called the *mantissa*, although this meaning should be restricted to the fractional part of a logarithm). The fraction f is bounded by $0.1 \leqslant f < 1$ and the exponent E is an integer chosen such that $f \times 10^E = N$, where N is the number to be represented. Thus our computer notation for 245.6 is 0.2456×10^3. Then, in our 6-decimal-digit computer we might see

    +   0   3   2   4   5   6

Usually, fractions are normalized, so that the fraction never has zero as a leading digit unless the number itself is zero. Also, the sign bit is the leftmost bit, as usual.

As shown here, there is a practical difficulty in representing numbers with negative exponents because there is only one sign bit. This is cleverly taken care of by "biasing" the exponent. An exponent is biased by adding a constant to it so that negative exponents become positive. In our decimal machine a suitable bias is 50 (the nearest round number to half the largest exponent that could be represented without the bias). With the bias, our number now appears as

    +   5   3   2   4   5   6

whereas $-3.79 \times 10^{-3} = -0.379 \times 10^{-2}$ would be

    -   4   8   3   7   9   0

The total number of different numbers that can be represented exactly on this machine can be found as follows:

- Fraction: $9 \times 10 \times 10 \times 10 = 9000$
- Exponent: $10 \times 10 = 100$
- Sign: 2

Thus we can represent $9000 \times 100 \times 2 + 1 = 1{,}800{,}001$ different numbers exactly. Note that we allow only nine choices for the first digit of the fraction because of the normalization requirement, and we add 1 for the exact representation of zero.

For a binary computer with 32-bit words, 24 of which are used for the fraction, we have:

- Fraction: 2^{23}
- Exponent: 2^{7}
- Sign: 2

For this case we have $2^{31} + 1 \cong 2.15 \ldots \times 10^{9}$ different numbers that can be exactly represented.

The finite size of the exponent limits the range (of size) of the representable numbers. In our decimal computer the smallest nonzero realized number would have an exponent (biased) of 00 and the fraction, 0.1000, which is 0.1×10^{-50} $= 10^{-51}$. The largest number would have 99 in the exponent and 0.9999 in the fraction, or $0.9999 \times 10^{49} \cong 10^{49}$. Thus we can represent a set of numbers X, where $10^{-51} < |X| < 10^{49}$, and zero. The 32-bit computer just described has a range of number size from about 2.7×10^{-39} to about 1.7×10^{38}, and, of course, zero.

During computation, if a number is produced that lies outside of the permitted range, there are several possible machine reactions. If the result is too small to be represented as a normalized floating-point number (*underflow*), it will often be converted into a machine zero without any warning, although some machines allow such an event to be flagged. The production of a result that is too large (*overflow*) may cause several kinds of reactions, ranging from an immediate halt to replacement of the result by a special character or representation (*infinity*) and a halt only when that representation is loaded into an arithmetic register. It is advisable for the user to determine exactly what the machine does in each circumstance.

One way to determine experimentally the number of bits in the binary floating-point fraction is to execute a small test routine:

1. Set $\epsilon = 1$.
2. Repeat $\epsilon = \epsilon/2$ until $1 + \epsilon = 1$.
3. $\epsilon = 2\epsilon$.

The final value of ϵ is the smallest amount that can be added to 1 and not "fall off the end of the register." It is sometimes referred to as the *machine epsilon*. For a 24-bit fraction, for example, $\epsilon = 2^{-23} = 1.192093 \times 10^{-7}$. If the machine has an arithmetic register longer than the numbers stored in memory, then the calculation of the machine epsilon must require storage and retrieval of epsilon at each stage of calculation.

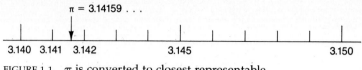

FIGURE 1.1 π is converted to closest representable number, 3.142, in 4-decimal-digit computer.

1.3   Representation of Data

Because of the limitation of the computer word length, not all numbers are representable. This often leads to the problem of introducing error into data values that are stored in the computer. For instance, our floating-point decimal machine with four digits in the fraction must represent π as 0.3142×10^1. On the number axis we can mark the numbers that are exactly representable, as in Figure 1.1. Any number that falls on the axis between these points can only be approximated in the machine.

We will agree on the following method of approximation, called *rounding*: For any number, choose the closest representable number. Because the representable numbers are spaced one machine epsilon apart (for numbers in the range from 1 to 9.999), we will never have more than one-half a machine epsilon relative error in representing any data number. In Chapter 2 we examine errors in more detail.

1.4   Operations in Floating Point

Standard arithmetic operations of addition, subtraction, multiplication, and division in floating point involve unpacking the floating-point numbers, doing the arithmetic operations with the fractional parts, normalizing, and packing the result. In larger and more sophisticated machines, these operations are carried out by the hardware, but in mini- and microcomputers the operations are usually carried out in software built into the high-level language processors. The following examples illustrate the basics of the operations in terms of the decimal computer we have been using before. The details are different for binary machines, but the principles are exactly the same.

Example 1.1   Divide 27.36 by 2.57.

These numbers are $+52.2736$ and $+51.2570$ in floating-point format; remember that these exponents are biased.

Unpack:	52	0.2736
	51	0.2570
Shift:	53	0.02736
Divide:	0.02736/0.2570 = 0.106459 . . .	
Normalize and round:	0.1065	
Subtract exponents and add bias:	53 − 51 + 50 = 52	
Pack:	+52.1065	(= 10.65)

The reason for the numerator shift is to prevent overflow in the divide hardware in case the numerator fraction is larger than that of the denominator; an automatic shift guarantees that the numerator fraction is always smaller than the denominator fraction.   ◻

Example 1.2   Divide 2.57 by 27.36.

Unpack:	51	0.2570
	52	0.2736
Shift numerator:	52	0.02570
Divide:	0.02570/0.2736 = 0.0939327 . . .	
Exponents:	52 − 52 + 50 = 50	
Normalize and pack:	+50.093932 . . . → +49.9393 (= 0.09393)	

Of course, in order to carry out these operations the hardware must accommodate more than the number of digits carried in the fraction. In binary computers the arithmetic registers used in floating-point operations usually have 2 or more extra bit positions to allow correct rounding.   ◻

Another example shows a different kind of shift problem that occurs in addition and subtraction.

Example 1.3   Add 159.1 and 3.652.

Unpack:	53	0.1591
	51	0.3652
Equalize exponents in order to align the decimal points:		
	53	0.1591
	53	0.003652
Add:	53	0.162752
Pack, normalize, and round:	53.1628	(= 162.8) ◻

These are examples of the kinds of operations that are done by the hardware (or software in smaller machines), usually in binary. You should work some of the problems at the end of the chapter in order to appreciate how the computer actually works.

1.5   Double-precision Mode

Many computer systems allow for an extended precision mode of representing numbers, often called *double precision*. Usually this mode is a function of software rather than hardware, although certain hardware features are helpful in making the system work easily. Double precision is a type declaration in some high-level languages, such as FORTRAN, so that the user is generally unaware of the details. Anyone doing serious computation should understand at least some of the principles involved.

The common way of accomplishing extra precision is to assign two consecutive words in memory to the storage of a single number, with the second word entirely devoted to storing additional bits of the floating-point fraction. For example, on a 32-bit computer with 24 bits used for the fraction (7+ decimal digits equivalent), the second word raises the length of the fraction from 24 to 56 bits (almost 17 decimal digits equivalent). In recent years, IBM and others have allowed a type declaration in FORTRAN that controls the precision:

        IMPLICIT REAL*8 (A-H, O-Z)   (8 bytes—64 bits)

or

        IMPLICIT REAL*4 (A-H, O-Z)   (4 bytes—32 bits)

Specific variables can be declared in violation of the implicit type declaration if the declaration follows the generic ones. For example, the usual FORTRAN declaration for double precision is

        DOUBLE PRECISION X, W, A

Users of Control Data Corporation (CDC) machines of 60-bit word length (almost 15 decimal digits in the fraction) need double precision less frequently. The CDC designers chose to structure the second word like the first, with 48 bits for the fraction and with an exponent scaled 2^{-48} with respect to the most significant word. Thus the two halves can be added in a regular double-length floating-point adder to produce properly a 96-bit fraction.

Double-precision arithmetic on most machines is generally expensive in terms of running time because of the many operations involved. Let us represent a double-precision number by

$$A_U + 10^{-t}A_L$$

where A_U is the more-significant fraction, A_L is the less-significant fraction, and 10^{-t} is the scale factor that relates the two fractions. Then

$$(A_U + 10^{-t}A_L)(B_U + 10^{-t}B_L) = A_UB_U + 10^{-t}(A_UB_L + A_LB_U) + 10^{-2t}A_LB_L$$

The product A_LB_L is usually not computed because it would affect only the round-off position of the less-significant half of the product fraction. Note that we have

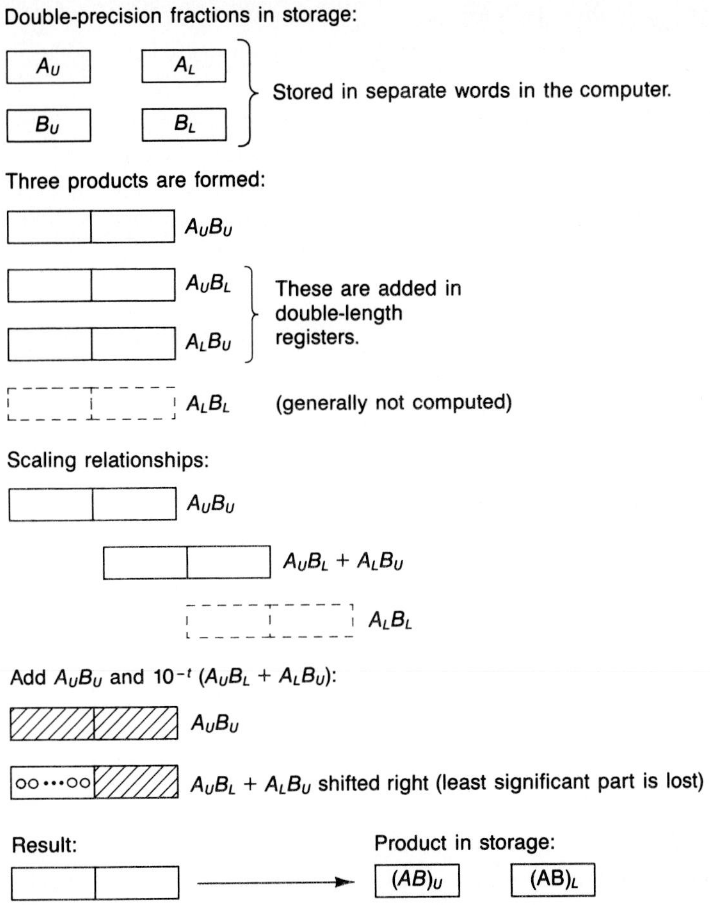

Double-precision fractions in storage:

A_U A_L } Stored in separate words in the computer.

B_U B_L

Three products are formed:

$A_U B_U$

$A_U B_L$ } These are added in
                double-length
$A_L B_U$ } registers.

$A_L B_L$ (generally not computed)

Scaling relationships:

$A_U B_U$

$A_U B_L + A_L B_U$

$A_L B_L$

Add $A_U B_U$ and $10^{-t}(A_U B_L + A_L B_U)$:

$A_U B_U$

oo $\cdots$ oo $A_U B_L + A_L B_U$ shifted right (least significant part is lost)

Result: Product in storage:

                    $\longrightarrow$ $(AB)_U$ $(AB)_L$

FIGURE 1.2   Double-precision multiplication.

to make three multiplications in place of one, and we have a number of additions that must be done carefully. This is best explained in Figure 1.2, where we see that there will be two additions with the possibility of addition overflow into the most-significant fraction of the product. That could cause the most-significant fraction to overflow, with the necessity of shifting part of the upper fraction back down into the lower. Needless to say, all of this takes time; generally, double precision runs perhaps five times as slowly as single precision in a computer.

1.6   Complex Mode

Some high-level languages, such as FORTRAN, support complex arithmetic involving numbers of the form $z = a + ib$, where i is the positive square root of -1. The common operations are:

1. Addition: $(a + ib) + (c + id)$ $= (a + c) + i(b + d)$

2. Subtraction: $(a + ib) - (c + id)$ $= (a - c) + i(b - d)$

3. Multiplication: $(a + ib) \times (c + id) = (ac - bd) + i(ad + bc)$

4. Division: $\dfrac{(a + ib)}{(c + id)}$ $= \dfrac{(ac + bd)}{(c^2 + d^2)} + \dfrac{i(bc - ad)}{(c^2 + d^2)}$

If complex mode is supported, the complex versions of the mathematical functions, such as square root, logarithm, sine, cosine, exponential, and arctangent, are usually provided in software.

Although these facilities are desirable to have, they can be provided by the user at the expense of extra programming. One way is to write subroutines to do the arithmetic and to provide the complex function calculations. The other way is to use complex analysis on the problem. For instance, if we want to solve the equation $f(z) = 0$, where $z = x + iy$, we can replace the original equation by

$$f(z) = u(x, y) + iv(x, y) = 0$$

where u and v are real functions, and then solve a pair of simultaneous equations $u(x, y) = 0$ and $v(x, y) = 0$. Reducing a complex expression to a pair of real expressions u and v involves the use of such common identities as

$$e^z = e^x(\cos y + i \sin y)$$

$$\log z = \text{Log}(x^2 + y^2)^{1/2} + i[\text{Tan}^{-1}\left(\frac{y}{x}\right) + 2n\pi] \qquad n = 0, 1, 2, \ldots$$

where Log means the principal-valued logarithm (which is the one found in ordinary tables). Anyone doing much of this kind of work would be well advised to review an introductory text in complex variables, such as Churchill and Brown [1984].

Exercises

1. Convert the following decimal numbers to octal:

   a. 35.75

   b. 87.875

   c. 39.3

   d. 859.76

   e. 0.0091

   f. 1,000,000

2. Convert the following decimal numbers to hexadecimal (base 16) with the symbol set {0, 1, 2, . . . , 9, A, B, C, D, E, F}:

   a. 32.75

   b. 126.375

   c. 43.3

   d. 927.37

   e. 0.0091

   f. 1,000,000

3. Convert the following octal numbers to decimal:

   a. 525.37   d. 336.3
   b. 63.73   e. 0.00723
   c. 1001.05   f. 1,000,000

4. Convert the following hexadecimal numbers to decimal:

   a. A3F.2B   d. 3F6.9
   b. 3C4.2A   e. 0.00F3A
   c. 100.0C   f. 1,000,000

5. Convert the following octal numbers to hexadecimal:

   a. 37.37   d. 365.2
   b. 235.06   e. 0.00723
   c. 1001.05   f. 1,000,000

6. Convert the following hexadecimal numbers to octal:

   a. A3F.2B   d. 3F2.9
   b. 37A.9   e. 0.00AFC
   c. 1001.05   f. 1,000,000

7. The IBM 650 (circa 1955) had a 10-digit decimal word with an extra sign bit. Floating point was often programmed using 8 digits for the fraction and 2 digits for the exponent. What would be the machine epsilon for this computer?

8. Show how the following decimal numbers would appear when stored in the 6-digit-decimal computer floating point as discussed in the text:

   a. 13927.35   d. 4.99994
   b. 6.023×10^{23} e. $-1.\overset{\cdot}{3}576$
   c. -0.00039658 f. -0.036527

9. Perform the following additions using the decimal computer discussed in the text (show the unpack, fraction operations, repack, and so on, using sign-magnitude representation):

   a. $154.9 + 2.357$ d. $5280 - 640.9$
   b. $129.12 - 3.652$ e. $56.98 - 55.97$
   c. $13245.7 + 23.9$ f. $89.89 + 98.98$

10. Perform the following multiplications using the decimal computer discussed in the text (see Exercise 9):

    a. 154.9×2.357 d. $5280 \times (-640.9)$
    b. $129.12 \times (-3.652)$ e. $56.98 \times (-55.97)$
    c. 13245.7×23.9 f. 89.89×98.98

11. Perform the following divisions using the decimal computer discussed in the text (see Exercise 9):

    a. $154.9/2.357$ d. $5280/(-640.9)$
    b. $129.12/(-3.652)$ e. $56.98/(-55.97)$
    c. $23.9/13245.7$ f. $89.89/98.98$

12. Suppose you have a 48-bit computer and you wish to design a floating-point representation that will have a range of about 10^{-38} to 10^{+38}. How many bits will be required for the exponent? Given that one bit is used for the sign, how many bits are in the floating-point fraction? What is the machine epsilon?

13. Repeat Exercise 12 for a 48-bit computer and a range of 10^{-308} to 10^{+308}.

14. Repeat Exercise 12 for a 64-bit computer with the same range as in Exercise 13.

15. Repeat Exercise 12 for a 60-bit computer with the same range as in Exercise 13.

16. For the computer that you are going to use in this course, find the machine epsilon experimentally. Write and run a program to do this for both single- and double-precision modes.

17. If you have a programmable pocket calculator, find the machine epsilon. (If it is a Texas Instruments calculator of a certain vintage, you may be surprised to find that the machine epsilon is smaller than expected for the number of digits displayed. Many TI calculators carry hidden digits beyond the number displayed in the register.)

18. Suppose you have a 6-decimal-digit-word computer with 4 digits for the fraction. Double precision adds 6 more digits (next whole word) to the fraction. Use this computer on the following computations:
    a. $23.67495812 \times 9371.374672$
    b. $129.8791639 + 2.520916471$
    c. $45.73091734/223.6187045$
    d. $0.002969372817 - 0.002749159284$

Problems

PROJECT PROBLEM 1.1

Write a PASCAL program that will allow the keyboard entry of a decimal number N and a base B and then convert N to a representation in base B. Limit the base to $2 \leqslant B \leqslant 16$ for practical reasons. Allow for a reasonable number of digits (e.g., 10) on each side of the radix point.

Suggestions

Separate the integer and fractional parts of N using, for instance, $T := INT(N)$ and $F := FRAC(N)$. Then apply the following conversion algorithms:

```
J := -1;
REPEAT
  F := B*F;
  A[J] := TRUNC(F);
  F := FRAC(F);
  J := J - 1
UNTIL J < smallest index for the array of coefficients
```

and

```
J := 0;
REPEAT
   T := T/B;
   A[J] := ROUND(B*FRAC(T));
   T := INT(T);
   J := J + 1
UNTIL ((J > largest index for array) OR (T = 0));
```

where A[. . .] is the array of digits for the new base system representation. Provide an error escape for the exit from the latter procedure in case T ≠ 0 when the exit is made; this is a case of overflow. A brute-force method can be constructed to convert the A[J] to characters in a new array C[J] (use a case statement, for example). Then the characters can be concatenated in a string with a period inserted in the proper place to serve as a radix point. The string can then be printed.

2

Errors and Mistakes

If numbers are the lifeblood of scientific and engineering computing, the operations on numbers that reduce the accuracy of the computations are the diseases against which we must guard. Error buildup, which may be subtle, can render virtually worthless the answers we seek. In this chapter we examine some of the ways in which this can happen. In order to show how it can happen, we begin with a simple problem: Find the roots of the quadratic equation

$$x^2 + 100{,}000x - 0.01 = 0$$

using the quadratic formula. Use of 48-bit arithmetic (about 11-decimal-digit floating-point fractions) yielded the real root pairs $(-1.0000000000 \times 10^5, 5.9604644775 \times 10^{-8})$ from the quadratic formula and $(-1.0000000000 \times 10^5, 1.0000000000 \times 10^{-7})$ by an improved method. Note that the error in the smaller root of the first answer is more than 40%. Why is the error in the smaller root so large? In the rest of this chapter we shall explore some of the problems of computing with approximate data and how to guard against such disastrous errors. At the end we shall see how to solve a quadratic equation correctly.

2.1 Definition of Error

The term *error*, as we shall use it, has a technical meaning. The true value of a number, such as π, can seldom be represented exactly in the computer, as we saw in Chapter 1. Rather, we must settle for an approximation, such as 3.14,

3.142, 3.1416, 3.14159, . . . , depending upon the length of the computer word. Also, numbers may represent measurements and thus may contain errors arising from the measuring instruments. We say that the error e is the true value of the number minus the approximate value of the number.

Error = true value − approximate value

Thus the error in the first approximation to π is

$$e = 3.14159 \ldots - 3.14 = 0.00159 \ldots$$

Another way of defining the error is to say that the true value equals the approximate value plus the error. Note that when the approximation is too small (for a positive number), the error is positive. If we divide the error by the true value, we have what is called the *relative error*:

$$\text{Relative error} = \frac{\text{error}}{\text{true value}}$$

$$= \frac{\text{true value} - \text{approximate value}}{\text{true value}}$$

In this example the relative error is

$$\frac{0.00159 \ldots}{3.14159 \ldots} = 0.000507 \ldots$$

which can also be expressed as a percentage. In this case, the relative error is $0.0507 \ldots \%$.

Both the error and the relative error can be useful, but the relative error is often more revealing. For example, an error of 1 mile in the distance to the sun gives a very small relative error (about 0.000001%). However, an error of 0.001 inch in measuring the thickness of a sheet of paper is relatively large (about 25%).

We must be careful to distinguish an error from a mistake. Common usage of the word *error* to mean mistake violates the technical meaning. For example, $\pi = 3.14$ has an error, but $\pi = 3.41$ is a mistake!

2.2   Sources of Error in Numbers

As we have emphasized before, the numbers used in computation almost always contain errors. Let us look at the sources of these errors so that we can see what to do about them.

2.2.1   Inaccurate Data

Much of the data used as input in scientific computation is of limited accuracy. Physical constants are seldom known to more than a few decimal digits of accuracy. Laboratory data typically contain errors ranging from 0.1% to several percent. (Time and frequency measurements are now done to accuracies such that the relative errors are on the order of the machine epsilon of many computers. Distance measurements can be made to high accuracy using optical means, but temperature or pressure transducers commonly used in laboratories are of limited accuracy.) The improvement of the accuracy of measurement is beyond the scope of this book. Nonetheless, the person using the computer must be aware of the accuracy of the data and not become beguiled by 10-digit answers based on 3-digit data.

2.2.2   Data Entry

Data entered from a keyboard, punched cards, or other input device are converted into binary form for storage in memory. This leads to additional error because a decimal fraction may not be representable exactly in binary. The decimal fraction 0.1, for example, converts to the nonterminating binary fraction 0.000110011001100 . . . , which must be cut off at some place. This adds to any errors in the input data. This may happen twice if a constant such as π is entered. Pi first must be rounded in decimal form and then converted to binary, which is again rounded.

2.2.3   Arithmetic Operations

Arithmetic operations on numbers in floating point generally degrade the accuracy. For example, if we divide an equation through by one of the coefficients in the process of solving the equation, we are likely to produce an error: $\frac{1}{3} = 0.333$ . . . cannot be represented exactly in either decimal or binary.

2.2.4   Formula Errors

Values of functions, such as sines, exponentials, and logarithms, are calculated using formulas that contain small errors. In the last chapter of the book we shall see how such approximations are devised and how to put bounds on the errors. Nonetheless, there will be small errors, which may be several times the size of round-off errors.

2.3   Round-off Errors

When numbers are shortened to fit into computer words of limited length, we say that the numbers contain *round-off errors*. The shortening of the numbers should be done properly, however, and simply dropping digits is not desirable. To round a number, such as π, to fit into a finite-length computer word (more properly, register), we should choose the remaining digits in order to minimize the error caused by the shortening. As we said in the first chapter, we should choose the closest representable number to the number being rounded. For instance, to round off π = 3.14159265359 . . . to 5 digits, we should use 3.1416, not 3.1415. The error in the first case is -0.0000073464 (to 10 places). In the second case it is $+0.0000926536$, a result that is over 12 times as great in magnitude. In other words, numbers should not be "chopped" (truncated). Unfortunately, there are many computers that truncate. The effect is less serious for those machines that have a long word length.

The algorithm for correctly rounding a positive number to n digits involves adding half the number base (1 in binary, 5 in decimal) to the $(n + 1)$st digit and then truncating. For example:

$$
\begin{array}{ll}
3.1415 \mid 9265359\ldots & \\
\underline{ \mid 5\longleftarrow} & \text{Add 5 in the first dropped place.} \\
3.1416 \mid 4265359\ldots & \\
3.1416 \mid \longleftarrow & \text{After truncation.}
\end{array}
$$

In effect, this also means that if the first dropped digit is 0 through 4, we drop the digit and its successors. If the first dropped digit is 5 through 9, we increase the last retained digit by 1 and then truncate.

Truncation, or chopping (as a method of rounding) always produces a positive error in the value of a positive rounded number. Properly rounded numbers tend to have randomly distributed positive and negative errors. Thus in the summation of a sequence of positive numbers, the properly rounded numbers tend toward a more accurate sum. Because the properly rounded numbers have errors that are both positive and negative, the errors tend to cancel. However, truncated numbers have consistently positive errors, which always add and never cancel. The error of the sum of properly rounded positive numbers, because of partial cancellation, tends to grow with the square root of the number of items. The error in the sum of truncated numbers, on the other hand, tends to grow proportionally to the number of data items. Proof of this can be found in Ralston and Rabinowitz [1978].

2.4 Significant Digits

We need to understand clearly what is meant by a significant digit. We say that a number that results from a measurement has as many significant digits as the accuracy of the measurement will allow. This can be illustrated and clarified by an example.

Figure 2.1 shows a scale with the edge of an object being measured. Obviously the object is between 2.4 and 2.5 units in length; in fact, it is between 2.47 and 2.48 units long. We can estimate from the scale that the length is 2.473 units long with some uncertainty. Experienced "guessers" (such a skill can be developed through experience) would say that the length is 2.473 units $\pm$ 0.0005 units, but for most people the accuracy of the guess is more like 2.473 $\pm$ 0.001 units. We

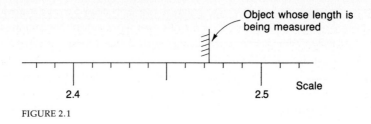

FIGURE 2.1

shall agree with the experienced estimators and accept ± 0.0005 as the uncertainty, with the understanding that this may be overoptimistic. The number of significant digits is 4, with a fractional uncertainty in the last digit. In fact, we can define the number of significant digits as the number of digits clearly correct plus 1 (for the digit that has an uncertainty on the order of half the base).

In most cases the number of significant digits is clear from the way in which the number is written. Most of the ambiguities involve zero.

Example 2.1

The distance to the sun is 93,000,000 *mi.* Clearly, this is meant to be a round number and not to be accurate to the nearest mile. It probably means 9.3×10^7 mi, which is accurate to 2 significant digits.   □

Example 2.2

1 *mi is equal to* 5280 *ft.* This is a definition, and the 5280 can be thought of as having unlimited accuracy—for example, has as many significant digits as needed: 5280.00000. . . . The same holds true in formulas where numbers such as 2 appear. For convenience, we sometimes call this kind of a number an *exact* number, meaning that it can be followed by a decimal point and as many zeros as desired in a computation.   □

Example 2.3

The thickness of the film is 0.0023 *cm.* In this case the zeros are used simply to locate the decimal point and are not significant, so that the number of significant digits is 2. Scientific notation clarifies the role of zeros: in this case the thickness can be written as 2.3×10^{-3}, which has two significant digits.   □

Example 2.4

The potential was 10.02 *V.* The zeros here are included between significant digits and are significant. Scientific notation does not eliminate these zeros: $10.02 = 1.002 \times 10^1$, so they are significant.   □

In general, any zeros not written when the number is written in scientific notation are not significant. Any zeros included between significant digits are themselves also significant. Trailing zeros, such as in the 93,000,000 mi, are open to question and may or may not be significant. The trailing digit in a number such as 0.00450 is supposed to be significant. The trailing zero should not be written unless it is significant (i.e., 0.00450 represents a number measured to be between 0.004495 and 0.004505).

2.5 Computation with Data

It is easy to lose sight of the fundamental limitations on the accuracy of the result of a computation due to the inaccuracy of the data used, especially when the computer presents us with 10- or 15-digit answers. As an example, we might wish to calculate the volume of a box that is 1.2 in. by 3.6 in. by 5.4 in. The computed result is 23.328 in^3. Unfortunately, each of the measurements has an uncertainty of $\pm$ 0.05 in., so that the result could be in error by more than 1 in^3. For example, suppose that the true values were all on the low side, that is, 1.15, 3.55, and 5.35; then the true volume would be as small as

$$1.15 \times 3.55 \times 5.35 = 21.841375$$

Likewise, all the true values could be on the high side, so that the true volume would be as large as

$$1.25 \times 3.65 \times 5.45 = 24.865625$$

From this we see that it is unreasonable to give the answer with more than 2 significant digits. It is possible that the relative error in the answer is greater than the relative error in any of the factors. (Check this by direct calculation.)

The use of extreme values of data in combinations that produce extreme values in computed answers is sometimes called *interval arithmetic*. Extreme values of the data are often called *range numbers*. While this kind of analysis is sometimes useful, it often gives very pessimistic error bounds in complicated calculations because it does not take error cancellation into account.

The following are some rules of thumb that are useful in dealing with calculations using approximate data:

1. A product or a quotient of two approximate numbers should have no more significant digits than the number of significant digits in the approximate number with the fewest significant digits.
2. The sum or difference of two approximate numbers should have no more significant digits to the right of the decimal point than the approximate number having the fewest significant digits to the right of the decimal point. For large numbers with few significant digits (e.g., 93,000,000 mi) the numbers should be expressed in scientific notation before this rule is applied.

3. In the subtraction of positive numbers of nearly equal size or in the addition of numbers of unlike sign but of nearly equal magnitude, there may be a loss of many significant digits.

In general these rules are limiting rules in the sense that there will be, for example, a small loss of accuracy in the multiplication or division process. The words *no more than* mean that one cannot gain accuracy, and in fact there usually will be a certain degradation of accuracy during computation.

To demonstrate Rule 1, let $x^* = x + \Delta x$ be the true value of a number, where x is its approximate value and where Δx is the error. Use corresponding notation for the number y^*. The true value x^*y^* of the product is given by

$$x^*y^* = (x + \Delta x)(y + \Delta y)$$
$$= xy + x\Delta y + y\Delta x + \Delta x\Delta y$$

and the error of the product is

$$x^*y^* - xy = x\Delta y + y\Delta x + \Delta x\Delta y$$

The relative error in the product is

$$\frac{x^*y^* - xy}{x^*y^*} = \frac{x\Delta y}{x^*y^*} + \frac{y\Delta x}{x^*y^*} + \frac{\Delta x\Delta y}{x^*y^*}$$

If the relative errors are small, the last term on the right-hand side is small with respect to the first two terms; also y/y^* and x/x^* are both very close to 1. Thus the relative error in the product is approximately equal to the sum of the relative errors in the factors:

$$\frac{x^*y^* - xy}{x^*y^*} \cong \frac{\Delta x}{x^*} + \frac{\Delta y}{y^*}$$

Note that Rule 1 holds: Whichever of x or y has the larger relative error dominates the relative error of the product. The case of division is left to the problems.

Rule 2 is best illustrated with an example.

Example 2.5

Suppose we wish to add 45.3 and 2.69. Both numbers have 3 significant digits and are therefore of approximately the same accuracy. But in forming the sum

45.3
2.69

we see that there is $\pm$ 5 uncertainty in the second digit to the right of the decimal point in the number 45.3, so that the sum could be as large as 48.04 or as small as 47.94. We have no right to deceive anyone by stating the sum as 47.99, which implies that the sum is a number that satisfies

47.985 < sum < 47.995

What we do is to add the numbers without rounding and then round the sum to match the number that has the fewest digits to the right of the decimal point. In this case, 47.99 rounds to 48.0, which says that the sum is a number satisfying

$$47.95 < \text{sum} < 48.05 \quad \square$$

Rule 3 covers a problem that is often troublesome and again is best shown with an example.

Example 2.6   Add 153.75 and -150.86.

$$\begin{array}{r} 153.75 \\ -150.86 \\ \hline 2.89 \end{array}$$

We started with numbers accurate to 5 significant digits and ended up with a 3-significant-digit answer! Unfortunately, there is no convenient way to detect the loss of significant digits during a computation. We must rely on our own skill and experience in setting up the problem to try to avoid such situations. $\square$

An example shows how a problem can be revised to avoid loss of accuracy on subtraction.

Example 2.7

Find the weight of a steel tube 16.00 ft in length with an outside diameter of 14.25 in. and an inside diameter of 13.25 in. if the specific weight of the particular steel is 459.0 lb/ft³.

The computer answer is 1101.520924 pounds. The cross-sectional area of the pipe is

$$A = \frac{\pi}{4}(14.25)^2 - \frac{\pi}{4}(13.25)^2$$

$$= 159.4849146 - 137.8864651 \quad \text{(4 significant digits)}$$

$$= 21.5984495 \rightarrow 21.6 \text{ inches}$$

Hence we see that the weight is accurate only to 3 significant digits and should be reported only as 1.10×10^3 pounds.

When we use interval arithmetic in the calculation, we find the following results:

$$W = \frac{\pi}{4}[(14.255)^2 - (13.245)^2]\left(\frac{16.005}{144}\right)(459.05) = 1113.00503 \text{ pounds}$$

$$W = \frac{\pi}{4}[(14.245)^2 - (13.255)^2]\left(\frac{15.995}{144}\right)(458.95) = 1090.04618 \text{ pounds}$$

Note that the possible values for the weight lie in a 2% range. The range for the data is on the order of only 0.1%.

This is a good example of the wrong way to gather data and solve the problem. A little algebra shows us that

$$W = \frac{\pi}{4}(D_o + D_i)(D_o - D_i)\frac{LW_s}{144} = \frac{\pi LW_s}{576}(D_o + D_i)(D_o - D_i)$$

where

- W is the weight in pounds.
- W_s is the specific weight in pounds per cubic foot.
- D_o, D_i are the diameters in inches.
- L is the length in feet.

The source of the trouble is $D_o - D_i$, which is twice the wall thickness of the pipe. A better way to produce an accurate estimate of the weight is to measure the wall thickness in several places in order to get an average thickness. Then use that, together with the diameter measurements, in the revised formula. An even more direct approach would be to weigh the pipe (provided a suitable scale is available)!   □

2.6   Truncation Errors and Errors in Formulas

Another fundamental source of errors in computation results from the termination of an infinite series expansion after a finite number of terms. We call this *truncation error*. For instance, we never can evaluate

$$\sin x = x - \frac{x^3}{3!} + \frac{x^5}{5!} - \frac{x^7}{7!} + \cdots$$

exactly except at the origin. The magnitude of the error depends upon the number of terms used and upon the value of x. We can see this in Figure 2.2, where we show the graphs of $\sin x$, x, $x - x^3/3!$, and $x - x^3/3! + x^5/5!$ on the same set of axes. Note that in each successive approximation to $\sin x$, the graph lies closer to that of $\sin x$ for a greater distance from the origin. By *closer* we mean that the absolute value of the difference between $\sin x$ and its approximation at a given value of x is smaller. This is characteristic behavior of truncation errors.

A way of estimating the size of the truncation error is by the use of Taylor's formula, or, as it is often called, Taylor's formula with remainder. We shall use the derivative form, which for this purpose is more useful.

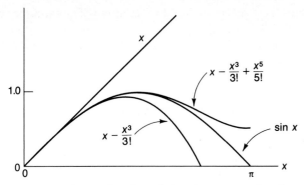

FIGURE 2.2   Approximation of sin x by truncated series.

Theorem 2.1

Let $a < x$, let the first n derivatives of $f(x)$ be continuous on $[a, x]$, and let the $(n + 1)$st derivative exist on (a, x). Then z exists, $a < z < x$, such that

$$f(x) = f(a) + f'(a)(x - a) + \frac{f''(a)(x - a)}{2!} + \cdots + \frac{f^{(n)}(x - a)^n}{n!} + \frac{f^{(n+1)}(z)(x - a)^{n+1}}{(n + 1)!} \quad \square$$

We can also assume that $x < a$ and have an analogous theorem with all the other inequalities reversed. Proofs of this theorem can be found in advanced calculus texts. The theorem does not tell us how to find z; it only guarantees that such a z does exist. In general, for a given a, the value of z will be different for each value of x.

Example 2.8

Find a bound on the error of $x - x^3/3!$ as an approximation to sin x at $x = \pi/2$.
   In this case we can write

$$\sin x = x - \frac{x^3}{3!} + \frac{f^{(4)}(z)(x - 0)^4}{4!}$$

Now $f^{(4)}(z) = \sin z$, so the remainder term is $x^4 \sin z/4!$. The largest possible value that sin z could have on the interval $(0, \pi/2)$ is 1, so we see that the remainder is surely bounded by $(\pi/2)^4(1)/4! = 0.25367$. The actual error is 0.07517, versus the 0.25367 remainder-formula bound. $\square$

We will find in succeeding chapters that this technique is very useful in establishing error estimates, albeit often very conservative ones. We will also find in other kinds of calculations that knowing the form of the remainder with respect to $(x - a)$ will be useful. You should be sure to work some of the problems of this kind at the end of the chapter in order to become thoroughly familiar with Taylor's formula.

2.7   Testing for Equality

We now need to look at the effects that errors can have on the way we program. One of the practical problems that we face is how to test for equality when we use floating-point arithmetic. As an example, suppose we try to execute the following piece of PASCAL code:

```
H := 1.0/3.0;
X := 0;
REPEAT X := X + H
UNTIL  X = 1.0;
```

The computer will not stop at X = 1.0 because the condition will never be true. If we use our 4-digit fraction decimal machine as an example, we see that H will be 0.3333, and X will go through the following sequence of values: 0, 0.3333, 0.6666, 0.9999, 1.332, . . . , none of which equals 1.0. We could replace = by => in the test, which would cause the computer to stop instead of going on forever, but the stop would occur at 1.332, not at the point where we want it to stop.

A way out of this dilemma is simply to replace the UNTIL condition by ABS (X - 1.0) < H/2.0, which will guarantee that we can now stop at the place we want to stop. In general, in dealing with floating-point arithmetic, we are forced to recognize the deficiencies in the computer arithmetic and replace strict equality with a kind of "fuzzy" equality, which says that if X is anywhere within a small band of uncertainty around a desired number, then we will accept X as having the desired value.

In many cases, however, we can avoid the problem by using integer arithmetic in the program-control structure:

```
H := 1.0/3.0;
X := 0;
FOR I := 1 TO 3 DO
      BEGIN
            X := X + H
      END
```

An even better procedure is to avoid the addition of increments and use the index to generate the values of X:

```
FOR I := 1 TO 3 DO
     BEGIN
            X := (I/3.0)
     END
```

For a general case of an interval from A to B with N subintervals and with X generated at the endpoints of all subintervals, we can use X := I * (B−A)/N for I := 0 to N. This method has the advantage that all points will be calculated with an error on the order of the machine epsilon or less, whereas with the increment method the Nth point can have as much error in its value as N times the machine epsilon, which may be a serious error if N is large.

2.8   How to Solve a Quadratic Equation

We now return to the example with which we began this chapter, where we solved a quadratic equation only to find one root seriously in error. The commonly stated solution of the quadratic equation $Ax^2 + Bx + C = 0$ provides good examples of the pitfalls of numerical computation when it is approached in a naive way. Unfortunately, the naive approach is built into countless numbers of computer programs.

One subtle but sometimes severe difficulty comes about because of the addition-subtraction loss of accuracy under certain conditions. Suppose $|4AC|$ $<< B$ and we attempt to use the usual formula for the roots:

```
X1 := (−B + SQRT(B*B − 4.0*A*C))/(2.0*A)
X2 := (−B − SQRT(B*B − 4.0*A*C))/(2.0*A)
```

In such a case the value of the radical expression, call it RAD, is very close to that of $|B|$. If B is positive, then −B + RAD is very close to zero and may well have very few significant digits, which will also be true for −B − RAD if B is negative. In either case there will be severe loss of accuracy in one of the roots. This is the case, of course, for the example given at the beginning of the chapter, where B was 100,000, A was 1, and C was --0.01. Furthermore, we may not know this for the case of a root-finding subroutine embedded in a large program that generates the coefficients A, B, and C automatically.

The proper way to proceed is to recognize that the equation may be rewritten as

$$x^2 + \left(\frac{B}{A}\right)x + \frac{C}{A} = 0, \qquad A \neq 0$$

which can be factored into the form

$$(x − R1)(x − R2) = 0$$

from which it is apparent that $R1 \times R2 = C/A$. We then calculate the first root $R1$ using

```
IF B > 0 THEN
    R1 := (-B - RAD)/(2.0*A)
    ELSE
    R1 := (-B + RAD)/(2.0*A)
```

and find the second root using

```
R2 := (C/A)/R1
```

Both roots are now of the same accuracy with no loss of significant digits from addition or subtraction. This is the method used to get the accurate answer in the example at the beginning of the chapter.

If the reason that $|4AC| \ll B$ is that A is very close to zero, then there is a possibility of overflow upon the division by $2.0*A$ in the calculation. It is a good precaution to test to see whether A is zero, in which case the equation can be solved for one root as a linear equation. If both A and B are zero, then there is no equation to solve at all. If A is not zero but if A/C gives a machine zero, then about the best we can do is to assume that one root is zero and solve for the other by setting $R2 = -C/B$. (Why?)

In writing such a subroutine we should remember that the quadratic equation can have complex roots and provide for finding these roots, which are complex conjugates. In this case subtraction is not a problem. An exercise is to produce such a procedure and exercise it with a driver program for a variety of different cases.

Exercises

1. Round the following numbers correctly to (a) 5 significant digits and (b) 4 significant digits. (Do *not* round to 5 digits and then to 4 digits!)

   a. 315.657   d. 7.765976
   b. 0.005972639   e. 6753.926
   c. 8.294348   f. 34.7295

2. Create an example to show why rounding one digit at a time can lead to an incorrect result. For example, you might be able to think of a sequence of 6 digits. If you round this sequence to 5 digits and then round that to 4 digits you will get a different (and wrong) result. However, if you round directly to 4 digits from 6 you will get the correct result.

3. Calculate the following volumes, using the approximate data given, and report the answer using the rules of thumb concerning accuracy and precision. Then calculate the range of volumes using interval arithmetic.

   a. Sphere, diameter = 1.62 in.
   b. Cylinder, diameter = 3.345 in., length = 1.37 ft

c. Box, 1.37 ft × 3.42 ft × 5.525 ft

d. Hollow cylinder, diameters 1.35 ft and 1.45 ft; length 4.645 ft

4. Perform the indicated arithmetic operations and give the answers correctly rounded under the assumption that the factors are stated with the number of significant digits appropriate to the measurement.

   a. 0.0032 + 1.29   **d.** 54.50/672.0

   b. 0.0032 × 1.29   **e.** 55.37(19.273 − 1.9752)

   c. 129.85 − 114.367   **f.** −0.0080 + 47.35

5. In the example of approximating the sine function using two terms of the series expansion, find the (approximate) value of z in the Taylor remainder formula to give the correct value of sine of $\pi/4$.

6. Find a bound on the error of estimating the values of the functions by the approximations given. Use the Taylor formula. Check the approximation by calculating the value of the function and the value of the approximation on your pocket calculator.

   a. $\sin x$, x, at $x = \pi/4$

   b. $\sin x$, $x - x^3/3! + x^5/5!$, at $x = \pi/2$

   c. $\cos x$, 1, at $x = \pi/6$

   d. e^x, $1 + x$, at $x = 0.1$

   e. e^x, $1 + x + x^2/2!$, at $x = 1.0$

7. Demonstrate Rule 1 for the case of a quotient of two numbers y/x using the same technique as that for multiplication. You will need to approximate $1/(x + \Delta x)$ as the series $(1/x)(1 - \Delta x/x + (\Delta x/x)^2 - \cdots)$ and then drop all but the first two terms.

8. For the computer that you are going to use in this course, determine the following characteristics of its arithmetic:

   a. Number of bits in floating-point single- and double-precision words.

   b. Number of bits in the floating-point fraction (both precisions).

   c. Range of size of numbers that can be represented.

   d. How are overflow and underflow handled in the computing system?

   e. Does the arithmetic round or truncate?

9. Write a PASCAL procedure or FORTRAN subroutine to solve a quadratic equation. The call to the procedure should resemble this:

      QUADRATIC (A, B, C, ROOTTYPE, X, Y) ;

where A, B, and C are the coefficients. Also, X and Y are the real roots or X and Y are the real and imaginary components of the complex conjugate roots or X is the single root in case A is zero. ROOTTYPE is a flag to distinguish among the cases: two real roots, a complex conjugate pair of roots, a linear equation with only one root, or a degenerate case with no roots (i.e., A = B = 0 regardless of the value of C).

    Write a driver program to allow keyboard input of A, B, and C in order to exercise the procedure for all possible cases.

10. Compute the following sums on your computer:

   a.
$$\sum_{i=1}^{100} (0.01)$$

   b.
$$\sum_{i=1}^{1000} (0.001)$$

c.
$$\sum_{i=1}^{10,000} (0.0001)$$

Subtract exactly 1 from each sum to show the round-off error effects.

11. Compute the following sums:

$$\sum_{i=1}^{32,767} \left(\frac{1}{i^n}\right) \quad \text{and} \quad \sum_{i=32,767}^{1} \left(\frac{1}{i^n}\right)$$

for $n = 1, 2,$ and 3. Are the values for the two sums done in the opposite direction the same? If not, why not?

3

Software

No less important than hardware, or the physical computer itself, is the software, or the programs that are run on the computer. The sophistication, reliability, and speed of machines have improved over the several decades, and the software has also changed. The earliest programs were single-purpose programs typically designed to produce tables of numbers. Today software encompasses a wide variety of applications programs (still often used to produce tables of numbers), plus many kinds of programs designed to help us write other programs (assemblers, interpreters, and compilers) as well as packages of subroutines that can be incorporated in the programs that we write. Various systems for producing graphical output replace the tables of numbers by graphs and views of surfaces that make the numbers much more meaningful. This is to say nothing of the many other kinds of software available to the scientist and engineer, such as spreadsheets, word processors, database management systems, and so-called expert systems, which attempt to recognize patterns in large masses of data.

3.1 High-level Languages

There are three different kinds of programs that assist us in writing other programs. The purpose of all three is the same—to translate programs written in some more-or-less high-level language to machine code. The first of these is called an assembler, and the language it uses is called assembly language. Assembly language is different for each machine or family of machines. Assembly language is close to machine language. The job of the assembler is the relatively easy one of replacing mnemonics (such as MOV, MUL, or ADD) by the numerical machine-operation code that we would find hard to remember and the symbolic names for variables (such as X, Y, TAXES) by the memory addresses where the variables are stored. Programming in assembly language is more tedious and more susceptible to mistakes than programming in higher-level languages. Thus it is done mainly for writing mathematical library functions, such as e^x or sin x, where the maximum efficiency of execution is desired, or for systems programming, where hardware control is a main consideration, speed is essential, and the hardware-control facilities are not provided in the high-level languages.

Interpreters (BASIC is one example) interpret the program statements into machine language at the time of execution of the statement. In contrast, compilers, such as FORTRAN and PASCAL, translate all the program statements into machine code before the program is executed. Interpreter programs thus tend to run many times more slowly than compiled programs because this interpretation must be done every time a program loop is executed. For this reason, some systems provide both a BASIC compiler and a BASIC interpreter. Once a BASIC interpreter program has been debugged and is running correctly, it can be compiled in order to save computer time if it is to be run many times with different data.

FORTRAN has been the most popular compiler language for scientific and engineering work. However, several newer languages enjoy growing popularity. Among these is PASCAL, which we will use in this book because it makes well-structured programs easier to write and to understand and because it is now widely used in teaching. FORTRAN has had several generations of extensions since its creation almost 30 years ago. For example, FORTRAN II, perhaps the first widespread version, did not have the logical IF statement, which was started in FORTRAN IV, or the IF-THEN-ELSE statement, which was started in FORTRAN 77.

Some of the strengths and weaknesses of various languages deserve mention. BASIC, for example, does not permit the establishment of libraries of already-compiled subroutines that can be linked and loaded with main programs, which is possible in FORTRAN and PASCAL. (A program loader is called after the main program is compiled. The loader locates and loads all separately compiled subprograms called for in the main program and then establishes the necessary links to the subprograms.) FORTRAN and PASCAL also have the feature that the formal parameters in functions and subroutines are "local" variables, so that a main program and a subroutine can use the same variable name for different purposes without conflict. In BASIC, no formal parameters are passed when a GOSUB is used. However, BASIC is easy to learn and easy to use for simple programs. For a small, simple program (no more than 100 lines of code), the use of a BASIC interpreter may provide answers more quickly than the use of one of the compilers.

PASCAL admits a wider variety of data types than does BASIC or FORTRAN, including pointers, sets, and records, which are very useful in many programming situations. However, it does not have type complex. FORTRAN is limited to integer, real, Boolean ("logical"), and (usually) double-precision and complex. String manipulations are very limited in FORTRAN. PASCAL has string and character types that allow string manipulations. BASIC systems also provide for type string. Some of the more elaborate BASIC languages have facilities for finding substrings and extracting blocks of characters from strings to form new strings.

Many FORTRAN compilers produce highly optimized machine code that runs faster than that from other compilers.

Some of the things that must be considered in making the choice of a system in which to do numerical work include the following:

- Is the system appropriate to the task? The problem may well be small enough to be done on a calculator. If there are no loops in the program, that is almost certain to be the case. Simple, one-loop problems often can be programmed and run on a programmable calculator in less time than on a computer.
- Does the language or system provide for integers and real numbers? Without this bare minimum we will have severe difficulty doing anything serious numerically. Complex arithmetic is also highly desirable for many numerical applications.
- Does the language or system provide for the separate compilation of subprograms so that a library of numerical procedures can be created and used in an easy way? This allows us to do complicated scientific or engineering calculations by writing a driver program that calls already-written software to do the numerical calculations.
- Are there good, already-existing subprograms or subroutines that are available in the literature or in the program library of the computer we wish to use?

One final word should be said about the importance of choosing to observe such standards as exist in defining a high-level language. Programs written in "standard" FORTRAN or PASCAL are as nearly machine independent as possible. This means that such programs can be moved from one machine to another with little or no change. This is an enormous advantage considering the amount of work involved in writing a large program.

3.2 Efficiency Versus Readability

Once a useful program has been written, it begins to have a life of its own. Modifications are often desired in order to accommodate new conditions or to guard against disasters that are observed to happen under certain conditions. Program maintenance, as it is usually called, is frequently the responsibility of someone other than the original programmer. For this reason it is desirable that the program be understandable and that the documentation be clear and complete. Programming tricks should be avoided, particularly those that depend on some idiosyncrasy of a particular manufacturer's hardware. (The same is true for software. Many programmers rued the day they started using the seven-character variable names permitted in CDC FORTRAN when they had to move the software to another machine, which allowed only the FORTRAN standard six-character names!)

Many large software projects have produced programs that have been used for 10 or more years. During that time the cost of maintenance often has exceeded the original cost of the program by several times. Everything that can be done initially to make the program maintainable later on is worth considering. More will be said about this later.

3.3   Writing Good Programs

The first and foremost step leading to a good program is to choose an appropriate algorithm. No amount of programming skill can compensate for the choice of a bad method. For now the only advice with respect to this matter is to keep your eyes open as you read the book in order to learn about good algorithms (and why some others are not so good).

The second consideration is writing a good program to implement the algorithm. By a good program we mean one that is correct (executes the intended algorithm and not some incorrect variant of it), is written with good style so that it is as readable and comprehensible as possible, and is well documented. Nothing is more discouraging than to deal with an old program that may be correct and readable, only to find that there is no way to tell whether the variable LENGTH is in inches, feet, or centimeters. The author's experience with such programs (some of which he wrote!) is that an hour of extra work on the documentation would have saved days of trying to reconstruct the answers to questions about the form of the input data by "deriving" the original formulas from the program. Likewise, units for output from the program should be printed by the computer. It is not a professional effort if units are entered on the output by hand.

Much computer output is used to support decisions to be made. A page full of numbers without explanation persuades the reader only of the incompetence of the person presenting the data and the results. Cryptic column headings are seldom helpful; AREA, SQ. FT. could be the area of anything. All tables should have descriptive headings and the columns should have headings with units. It should always be clear from the table headings, column headings, and row headings what is being tabulated.

Initial data and the values of program constants should also be printed out, with units. It is always important to know the values of parameters used.

3.3.1   Programming Style

It used to be that any program that got a job done was a "good" program. Today we demand much more from programmers. As we mentioned before, the ease of understanding a program greatly enhances its maintainability. For this reason we should strive to develop a good programming style. Fine resources for learning good programming style are *Elements of Programming Style* [Kernighan and

Plauger, 1974] and *Professional Pascal* [Ledgard, 1986]. Some of the components of good style include the following:

- *Top-down design*. In top-down design, the algorithm is studied at length to see what its components are. Then a grand design is developed before any code is written. It is analogous to the procedure in designing an airport. The first decisions involve questions concerning the amount of air traffic expected now and in the future. This dictates the amount of land necessary for landing strips and for terminal facilities and access roads. Then the land must be secured. Only after that can detailed design begin on runways, terminals, roads, and parking. Unfortunately, we sometimes have a tendency to want to dive into the fascinating details before we have settled the major issues.

- *Modularity*. Once the overall design has been decided, the blocks of the program should be laid out. What does a particular block do? What are its inputs? What are its outputs? How should blocks communicate with each other?

- *Attention to detail*. In writing the code, it is desirable to use such simple and effective devices as indenting the lines of code to show where program controls begin and end. This makes it much easier to keep track of the depth of loop structures and the depth of IFs contained within IFs. Liberal use of comments also makes it easier to understand the program.

- *Block structures*. Many programs suffer from the indiscriminate use of unconditional transfers (GO TO) to alleviate problems of careless design. The resulting "plate-of-spaghetti" programs are difficult to understand. Also, program modifications are unpleasant and sometimes disastrous. Often such spaghetti programs are the result of spaghetti thinking and betray a lack of a clear understanding of the program before the programmer began to write the program. Likewise, conditional exits from the interiors of loops are to be avoided. They usually can be in PASCAL if serious thought is given to how a loop is structured. Perhaps the only legitimate case of exit from the middle of a loop is the disaster exit (unacceptable data, for example), which terminates the program. This kind of exit does not complicate the logic of the program and allows an exit to a program block containing error messages to be printed and a halt. (Obviously a halt could also be done inside the loop, but it adds to the clutter. If it is understood that such exits are emergency exits only, then it makes a cleaner program.)

- *No tricks*. Avoid programming tricks, obscure code, or special features of a particular machine.

3.3.2 Documentation

Programs of substantial size are never self-contained. Documentation, in addition to internal program comments, is essential. Much commercial software is protected, so that the user cannot obtain a listing of the program and its internal comments. Hence external documentation is required.

For elaborate programs or sets of library subroutines, the documentation may run to several volumes. For simpler programs the documentation may be as simple as one or more typewritten pages. The nature and extent of the documentation varies with the kind of user.

The first level might be the *user guide*, which tells how to run the program or how to attach the subroutine to the main program. Usually this kind of document explains the data input and output and how to make it work. It often contains sample runs. In the case of an elaborate scientific subroutine package, such as IMSL (International Mathematical and Statistical Library), the documentation required is relatively minimal because the users are relatively sophisticated. A one- or two-page description of the subroutine-call parameters, the input data format, the method (with references to the literature) and its limitations, and a small example are all that are needed. Nonetheless, the manual consists of three loose-leaf volumes because of the number of programs. In contrast, a word processing program to run on a home computer typically includes a large reference manual, a self-teaching manual, and perhaps some demonstration or tutorial programs so that unsophisticated users can learn how to use something relatively sophisticated.

Documentation of the design or of the program itself is another matter. The source program and its internal documentation (comments) are usually available to the person who is interested in the design of the program or in its maintenance. The internal documentation (comments) is usually insufficient for understanding a large program, which makes external documentation essential. If you invest more than a few hours in writing a program, then you owe it to yourself (and to your employer) to spend at least as much time on documentation as you spent on writing and debugging the program.

3.3.3 Testing

How do we know that a program works? The answer is that seldom can we prove correctness. There are logical proofs of correctness that have been applied to small programs, but the labor of applying these methods to a substantial program is prohibitive.

Usually we take the pragmatic approach and test the program against a variety of problems whose answers we know. For instance, if we write a subroutine to find

$$I = \int_b^a f(x)\, dx$$

given a, b, and $f(x)$, then we can test the program for a variety of $f(x)$ and over many intervals $[a, b]$ for which we know the values of I exactly. After many such tests without failure (meaning that $|I^* - I|$ is sufficiently small, where I^* is the true value), we accept the program with reasonable confidence that it is correct. Most numerical analysts know some pathological problems to pose as test cases. If the method succeeds for these, we have even more confidence. It is a good idea not to limit your test functions to polynomials. Many methods are derived using an interpolating polynomial as a replacement for $f(x)$ and performing the mathematical operations on the polynomial. The results may be misleading, indicating more accuracy than actually exists for most functions.

3.3.4 Special Cases

If a program is written with sufficient modularity, it is easy to introduce special cases that may have known solutions. For example, suppose we wish to solve the differential equations for the flight of a cannonball through the air. In addition to the gravitational force, we have an air-drag force and a pseudoforce (Coriolis force), which is really the effect of the earth's rotation while the ball is in flight. We shall neglect the Coriolis force here, as well as any aerodynamic effects due to the spinning of the ball.

The magnitude of the drag force is given by $1/2 A\rho C_D V^2$, where A is the cross-sectional area of the ball, ρ is the air density, V is the velocity, and C_D is the drag coefficient, dependent on the mach number corresponding to V. The mach number is the ratio of V to the speed of sound in air of that density and temperature. Various models can be used to describe the density and temperature variations with altitude; another model can be used to describe the behavior of the speed of sound as a function of density and temperature.

A modular program consists of a main (*driver*) program with responsibilities for input of data, initializing variables, producing output, and calling upon the other modules as necessary. The drag force can be obtained from a function that returns the value of the force given V as a parameter. However, it must call upon another function to calculate C_D. This, in turn, must call on a still-different function to compute the mach number. That this entire process will work the first time is too much to expect! A very simple first step is to make the drag force function return the value zero whenever it is called. This reduces the problem to a very simple one:

$$\frac{md^2x}{dt^2} = 0$$

$$\frac{md^2y}{dt^2} = -g$$

where m is the mass and g is the acceleration of gravity. The solution is

$$x = x_0 + v_{x_0} t$$

$$y = -(\frac{g}{2m})t^2 + v_{y_0} t + y_0$$

If the differential-equation solver produces values very close to those obtained from these simplified equations you are in luck. If not, then you must find out why not. The troubles are either in the differential-equation solver or in the main program. If the differential-equation solver is known to be correct, then the main program is wrong in some aspect. (Are the parameters of the call to the differential-equation solver subroutine correct? Are the differential equations correct? Is there confusion concerning units? Note that the mass must be in slugs (1 slug = 32.2 lb mass, or lbm) in the British system.)

The next step is to introduce nonzero drag, perhaps first by using a constant drag coefficient. The equations are now

$$m\frac{dv_x}{dt} = -\frac{1}{2}A\rho C_D V^2 \cos\theta = -\frac{1}{2}A\rho C_D V V_x$$

$$m\frac{dv_y}{dt} = -mg - \frac{1}{2}A\rho C_D V^2 \sin\theta = -mg - \frac{1}{2}A\rho C_D V V_y$$

If we temporarily eliminate gravity, we can divide one equation by the other in order to obtain

$$\frac{dv_y}{dv_x} = \frac{V_y}{V_x}$$

or

$$V_y = kV_x$$

where k is some constant. With gravity eliminated, does the velocity behave this way? If not, there is a problem.

The higher levels of sophistication such as the variation of density with altitude, can now be added. As these complications are introduced, tests of "reasonableness" must be applied. For instance, the air density decreases with altitude. If the range of the cannonball does not increase because of the introduction of a density model in which density decreases with altitude, then there must be something wrong with the density calculation.

A final word on procedures or subroutines: Any new procedure or subroutine should not be first tested in the main program. Rather, it should be exercised with a simple driver program that prints out the values returned by the subprogram. This serves as a check before the subroutine is incorporated into the main program.

3.4   Use of Libraries

Libraries of mathematical subroutines and functions provide powerful tools for the scientific or engineering programmer. Unfortunately, some of the tools that have been offered by computer manufacturers have not been of the best quality. Many years ago the author encountered a cosine function that gave a value of zero for an argument of zero. For an argument very slightly away from zero, the correct cosine value was given. It was apparent that the programmer had included a test for zero and had chosen the wrong return value. Fortunately, such egregious errors are not common. Even so, the quality of manufacturer-supplied numerical software has not been outstanding.

More recently numerical analysts have collaborated in efforts to produce subroutine libraries of excellent quality and of wide variety. Substantial amounts of computer time have gone into running test cases to ensure that subroutines perform as intended. Also, there are excellent programs that have become part of the public domain. A few of these have been translated into PASCAL for use in this book.

3.4.1   IMSL

IMSL, Inc., is an organization that distributes the IMSL library, a very large collection of (FORTRAN) subroutines. Versions are provided for all the common mainframe computers as well as for several "superminis." More recently, a small version for the IBM PC and compatibles has become available. The selection and implementation of these subroutines is under the supervision of an advisory board of distinguished numerical analysts. New versions of the library are issued periodically. The library is available on a lease basis.

The topics covered range from statistics through linear algebra, integration, and solution of differential equations to special functions. The large number of subroutines (many hundreds) results from an attempt to provide for a wide variety of special cases. For example, linear-equation solvers are provided for symmetric matrices (clever storage schemes can save almost half of the storage space), banded matrices (only the elements lying within the bandwidth along the diagonal of the matrix are nonzero), complex matrices, general nonsymmetric matrices, and the like. The algorithms used are those considered to be the best known for the purpose, with a good selection of algorithms available for handling special cases. The quality of work in implementing them is first rate.

3.4.2   LINPACK

LINPACK is a collection of subroutines for the solution of systems of linear equations. These were largely generated at Argonne National Laboratory with the aid of interested persons from a variety of other organizations. These are excellent programs but limited to this one area. A user's guide, with FORTRAN listings of the source programs, is available from the Society for Industrial and Applied Mathematics (SIAM). Computer tapes of the source code are available from Argonne and from IMSL.

3.4.3   Other Libraries

There are several other libraries of good programs available through several different sources. Some of these are listed next. Many of them have resulted from the collaborative efforts of numerical analysts in a variety of different agencies, such as the Argonne National Laboratory and the National Center for Atmospheric Research, and represent good algorithms and good software.

EISPACK   EISPACK is a collection of subroutines for solving eigenvalue problems. The programs are available from IMSL, and documentation is available from Springer-Verlag.

FUNPACK   Special functions (Bessel functions, exponential integral, and the like) are available in the FUNPACK package of subroutines available from the National Energy Software Center, Argonne National Laboratory.

PPPACK   PPPACK is a collection of polynomial interpolation and spline interpolation subroutines available through IMSL.

Another source of good numerical algorithms is the Collected Algorithms of the Association for Computing Machinery (ACM), often referred to as CACM. These have been accumulated since about 1960. Early rules of the ACM limited the publication language to ALGOL, but since about 1970 this restriction has been relaxed. Beginning in 1975 the ACM made the algorithms available on tape or cards from the ACM Algorithms Distribution Service for a modest charge.

Problems

PROJECT PROBLEM 3.1

Create a program to solve quadratic equations. The program should consist of a driver program for input and output with a separately compiled quadratic-equation solver employing the principles discussed at the end of Chapter 2. Provide for complex roots as well as real roots, and provide a graceful way of handling input data disasters (e.g., $a = b = c = 0$).

If your system does not permit separate compilation of subprograms (e.g., Turbo PASCAL), then you will need to use "include" files to accomplish the same thing—the establishment of a subprogram library.

Provide a user's guide. Be sure to include comments in your source code. Describe the required input data, the furnished output data, and how errors are handled.

When you are satisfied with the program and its documentation, exchange programs with a fellow student. Each of you now should try to operate the other person's program using only the user's guide. When you have been able to make the other person's program run and have tested it out on a variety of cases, note any idiosyncracies or deficiencies on a sheet of paper. Include comments on ease of use, clarity, and so on, and submit the program and your comments to your instructor.

PROJECT PROBLEM 3.2

Investigate Cardan's formulas for the roots of a cubic equation and Ferrari's method for finding the roots of a quartic equation.

Expand Project Problem 3.1 to include roots of equations up through fourth degree. (No such closed procedures are available for equations of degree higher than fourth. In Chapter 5 we will discuss iterative methods for solving such equations.)

4

Systems of Linear Algebraic Equations

4.1 Introduction

Systems of linear equations arise in a variety of problems, such as the flow of fluids in pipe networks, flow of electric currents in linear electrical networks, least-squares statistical calculations, and stress calculations in structures. Numerical solutions of partial differential equations by finite-difference or finite-element methods lead to linear systems of equations that are often of very large size. Let us look at a simple example from statics to show how a problem involving simultaneous linear equations can arise.

Figure 4.1 shows a pin-jointed plane structure, which could be the side of a small truss bridge. By *pin-jointed* we mean that a member is free to rotate about a joint unless the other end of the member is constrained. (Two wooden strips fastened together by a single nail are pin-jointed.) The assumption of pin-jointedness greatly simplifies the analysis of the structure because we do not have to take into account any torques transmitted by members. As a consequence the members can have forces only along their lengths, either tension or compression. The truss is supported at the left end on a pin joint and is free to slide at the right end, as indicated by a roller. A single external force is applied at the center. This force contains the weight of the truss as well as the load on the truss. Reaction forces R_1 and R_2 occur at the left end, along with a single reaction force at the right end. (At the left end we need two forces because F does not necessarily have to be vertical.)

A principle of statics is that the vector sum of all the external forces (F and R's) on the truss must be zero. If not, then Newton's second law of motion

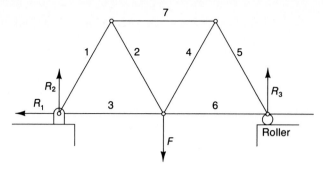

FIGURE 4.1

requires that the truss accelerate. This principle also applies at each of the joints. There are 5 joints and there are 10 forces, so that if we resolve each of the member forces into horizontal and vertical components, then we can find 10 equations from which to solve for the 10 unknowns.

First we arbitrarily assume that all forces in members are compression forces. If that turns out not to be the case for any member, then the value comes out negative. Let us apply the static equilibrium condition at the upper left joint, using the usual sign convention of positive for components up or to the right:

$$F_1 \cos 60° + F_2 \cos 120° - F_7 = 0 \text{ (horizontal)}$$
$$F_1 \sin 60° + F_2 \sin 120° = 0 \text{ (vertical)}$$

or

$$0.500F_1 - 0.500F_2 - 1.000F_7 = 0$$
$$0.866F_1 + 0.866F_2 = 0$$

where only three significant digits are shown. You are encouraged to work out the rest of the equations to show that the system is as follows:

$$0.500F_1 - 0.500F_2 - 1.000F_7 = 0$$
$$0.866F_1 + 0.866F_2 = 0$$
$$0.500F_4 - 0.500F_5 + 1.000F_7 = 0$$
$$0.866F_4 + 0.866F_5 = 0$$
$$-0.500F_1 - 1.000F_3 - 1.000R_1 = 0$$
$$-0.866F_1 + 1.000R_2 = 0$$
$$0.500F_2 + 1.000F_3 - 0.500F_4 - 1.000F_6 = 0$$
$$-0.866F_2 - 0.866F_4 = F$$
$$0.500F_5 + 1.000F_6 = 0$$
$$-0.866F_5 + 1.000R_3 = 0$$

These equations can be put into matrix form $Ax = b$, where A is the matrix

$$
\begin{bmatrix}
0.50 & -0.50 & 0.00 & 0.00 & 0.00 & 0.00 & -1.00 & 0.00 & 0.00 & 0.00 \\
0.87 & 0.87 & 0.00 & 0.00 & 0.00 & 0.00 & 0.00 & 0.00 & 0.00 & 0.00 \\
0.00 & 0.00 & 0.00 & 0.50 & -0.50 & 0.00 & 1.00 & 0.00 & 0.00 & 0.00 \\
0.00 & 0.00 & 0.00 & 0.87 & 0.87 & 0.00 & 0.00 & 0.00 & 0.00 & 0.00 \\
-0.50 & 0.00 & -1.00 & 0.00 & 0.00 & 0.00 & 0.00 & -1.00 & 0.00 & 0.00 \\
-0.87 & 0.00 & 0.00 & 0.00 & 0.00 & 0.00 & 0.00 & 0.00 & 1.00 & 0.00 \\
0.00 & 0.50 & 1.00 & -0.50 & 0.00 & -1.00 & 0.00 & 0.00 & 0.00 & 0.00 \\
0.00 & -0.87 & 0.00 & -0.87 & 0.00 & 0.00 & 0.00 & 0.00 & 0.00 & 0.00 \\
0.00 & 0.00 & 0.00 & 0.00 & 0.50 & 1.00 & 0.00 & 0.00 & 0.00 & 0.00 \\
0.00 & 0.00 & 0.00 & 0.00 & -0.87 & 0.00 & 0.00 & 0.00 & 0.00 & 1.00
\end{bmatrix}
$$

x is a vector of forces

$$
\begin{bmatrix}
F_1 \\
F_2 \\
\cdot \\
\cdot \\
\cdot \\
R_3
\end{bmatrix}
$$

and b is the vector

$$
\begin{bmatrix}
0 \\
0 \\
\cdot \\
\cdot \\
\cdot \\
0 \\
F \\
0 \\
0
\end{bmatrix}
$$

The solution for this set of equations with $F = 1000$ lbf can be found using the equation solver discussed later in this chapter. Correct to 3 significant digits, the forces found by the solver are

$$F_1 = 577 \text{ lbf}$$
$$F_2 = -577 \text{ lbf} \quad \text{(tension)}$$
$$F_3 = -287 \text{ lbf} \quad \text{(tension)}$$
$$F_4 = -577 \text{ lbf} \quad \text{(tension)}$$
$$F_5 = 577 \text{ lbf}$$
$$F_6 = -287 \text{ lbf} \quad \text{(tension)}$$
$$F_7 = 577 \text{ lbf}$$
$$R_1 = 0 \text{ lbf}$$

$$R_2 = \quad 500 \text{ lbf}$$
$$R_3 = \quad 500 \text{ lbf}$$

In order to learn how an equation solver such as the one used to solve this problem works, we need to review some elementary matrix algebra. You should be familiar with this material.

4.2   Review of Matrix Algebra

A matrix is a rectangular array of numbers. The numbers may be integers, real numbers, or complex numbers. We shall use the notation

$$A = \begin{bmatrix} a_{11} & a_{12} & \cdots & a_{1n} \\ a_{21} & a_{22} & \cdots & a_{2n} \\ \vdots & \vdots & & \vdots \\ a_{m1} & a_{m2} & \cdots & a_{mn} \end{bmatrix} = (a_{ij})$$

to designate an $m \times n$ matrix, one containing m rows and n columns. The first subscript designates the row, and the second subscript designates the column. Much of the time we shall deal either with square matrices ($m = n$) or with rectangular matrices in which one of the dimensions is 1. The latter can be of two varieties: a row vector ($1 \times n$) or a column vector ($m \times 1$).

A matrix (or vector) in which all elements are zero is called a *zero matrix* (or vector). A square matrix (a_{ij}) in which all elements are zero except those for which $i = j$ is called a *diagonal matrix*. A special case of a diagonal matrix is the identity matrix, in which all the diagonal elements are 1. If all elements are 0 except those for which $|i - j| < 2$, the matrix is called *tridiagonal*. A tridiagonal matrix (nonzero elements along the main diagonal, the superdiagonal, and the subdiagonal) is an example of a band matrix, which is defined as a matrix in which $a_{ij} = 0$ for all i, j such that $m \leq |i - j| < n$, where n is the order (number of rows and columns) of the matrix.

4.2.1   Unary Operations

Unary operations are those involving a single matrix. One of these is negation: $-A = (-a_{ij})$. For example,

$$-\begin{bmatrix} 1 & -3 \\ 2 & 4 \end{bmatrix} = \begin{bmatrix} -1 & 3 \\ -2 & -4 \end{bmatrix}$$

Another is the operation of transposition, indicated by the superscript T: $A^T = (a_{ij})^T = (a_{ji})$.

Example 4.1

$$\begin{bmatrix} 1 & 2 \\ 3 & -4 \end{bmatrix}^T = \begin{bmatrix} 1 & 3 \\ 2 & -4 \end{bmatrix}$$

$$\begin{bmatrix} 1 & -2 & 4 \\ 3 & 3 & 9 \end{bmatrix}^T = \begin{bmatrix} 1 & 3 \\ -2 & 3 \\ 4 & 9 \end{bmatrix}$$

$$\begin{bmatrix} 1 & 2 & 5 \end{bmatrix}^T = \begin{bmatrix} 1 \\ 2 \\ 5 \end{bmatrix}$$

$$\begin{bmatrix} 5 \\ -3 \\ 6 \\ 4 \end{bmatrix}^T = \begin{bmatrix} 5 & -3 & 6 & 4 \end{bmatrix} \quad \square$$

Note that for square matrices, the elements along the main diagonal remain fixed under transposition. Elements off the main diagonal exchange positions symmetrically with respect to the main diagonal.

A matrix A for which $A = A^T$ is called *symmetric*.

4.2.2 Binary Operations

Various algebraic operations are defined that involve two matrices as operands. First we must define equality of matrices. Matrix A equals matrix B if A and B have the same dimensions and if $a_{ij} = b_{ij}$ for all i and j. We then can write $A = B$.

Addition and Subtraction If A, B, and C are all matrices of the same dimensions, then we define matrix addition $C = A + B$ by

$$c_{ij} = a_{ij} + b_{ij}$$

for all i and j. We simply add corresponding elements to obtain the matrix sum:

$$\begin{bmatrix} 0 & 1 & 3 \\ 2 & -1 & 5 \\ 1 & 2 & 9 \end{bmatrix} + \begin{bmatrix} 3 & -2 & 4 \\ 5 & 1 & 7 \\ -2 & 1 & 4 \end{bmatrix} = \begin{bmatrix} 3 & -1 & 7 \\ 7 & 0 & 12 \\ -1 & 3 & 13 \end{bmatrix}$$

The matrix difference is the same except elements are subtracted instead of added.

Multiplication First we define multiplication of a matrix by a scalar as multiplication of every element in the matrix by the scalar:

$$mA = (ma_{ij})$$

For example,

$$2\begin{bmatrix} 1 & 3 \\ -2 & 4 \end{bmatrix} = \begin{bmatrix} 2 & 6 \\ -4 & 8 \end{bmatrix}$$

In order to have a meaningful product of two matrices, the matrices must be conformable. This means that in the product $C = AB$, the number of columns of A must be the same as the number of rows of B. The dimensions of C are as follows: the number of rows of C is the same as the number of rows of A and the number of columns of C is the same as the number of columns of B. Then

$$c_{ij} = \sum_k a_{ik}b_{kj}$$

where the summation on k ranges over the number of columns of A (or rows of B).

Example 4.2

$$\begin{bmatrix} 1 & 2 \\ 4 & 6 \end{bmatrix}\begin{bmatrix} 3 & 2 \\ -1 & 5 \end{bmatrix} = \begin{bmatrix} 1\times3 + 2\times(-1) & 1\times2 + 2\times5 \\ 4\times3 + 6\times(-1) & 4\times2 + 6\times5 \end{bmatrix}$$

$$= \begin{bmatrix} 1 & 12 \\ 6 & 38 \end{bmatrix}$$

$$\begin{bmatrix} 1 & 3 & 2 \\ 2 & -1 & 4 \end{bmatrix}\begin{bmatrix} 1 & 3 & 2 \\ 2 & 1 & -1 \\ -2 & 4 & 2 \end{bmatrix} = \begin{bmatrix} 3 & 14 & 3 \\ -8 & 21 & 13 \end{bmatrix}$$

$$\begin{bmatrix} 1 & 2 & -1 \end{bmatrix}\begin{bmatrix} 2 \\ -3 \\ 4 \end{bmatrix} = 2 - 6 - 4 = -8 \quad \square$$

Note that the last example is the ordinary scalar, or *dot*, product of two vectors. We will follow the convention that a column vector is the *standard* form of a vector and that a scalar product is designated by a^Tb, where a^T is a row vector.

Matrix multiplication is not commutative in general. For example,

$$\begin{bmatrix} 1 & 2 \\ 3 & 4 \end{bmatrix}\begin{bmatrix} 2 & -3 \\ 2 & 1 \end{bmatrix} = \begin{bmatrix} 6 & -1 \\ 14 & -5 \end{bmatrix}$$

However,

$$\begin{bmatrix} 2 & -3 \\ 2 & 1 \end{bmatrix}\begin{bmatrix} 1 & 2 \\ 3 & 4 \end{bmatrix} = \begin{bmatrix} -7 & -8 \\ 5 & 8 \end{bmatrix}$$

The transpose of the product of two matrices is the product of the transposes in reverse order:

$$(AB)^T = B^T A^T$$

A formal proof is found in most texts on linear algebra.

4.2.3 Inverse of a Matrix

Division of matrices is not defined, but there is a matrix operation analogous to that of finding the reciprocal of a number. First we must deal with properties of identity matrices, which we previously defined as square matrices with 1s along the main diagonal and 0s elsewhere. They are called *identity* matrices (usually designated as I or sometimes as I_n if the order n needs to be designated explicitly) because of the property that $IA = AI = A$, where A and I are conformable.

We define the inverse A^{-1} of a (square) matrix A as the matrix (if it exists) for which $AA^{-1} = A^{-1}A = I$. In particular, $I^{-1} = I$. If A^{-1} fails to exist, we say that A is singular. The following statements can be shown to be equivalent:

- A is singular.
- The determinant of A is zero.
- The rows (or columns) of A, considered as vectors, are not independent.
- The solution of $Ax = b$, for b nonzero, does not exist or is not unique.

4.2.4 Vector and Matrix Norms

A norm is a measure of the size of a vector or matrix. For vectors we often associate the idea of Euclidean length with the norm. Thus a vector (3, 5, 2) has a length $(3^2 + 5^2 + 2^2)^{1/2} = (38)^{1/2}$. Certain properties are required of a norm with which the Euclidean length is consistent. For a vector $\mathbf{v}$ with norm $\|\mathbf{v}\|$, these properties are:

1. $0 \leqslant \|\mathbf{v}\|$ for all $\mathbf{v}$; $\|\mathbf{v}\| = 0$ if and only if $\mathbf{v}$ is a zero vector.
2. $\|c\mathbf{v}\| = |c| \cdot \|\mathbf{v}\|$ for any real or complex number c.
3. $\|\mathbf{u} + \mathbf{v}\| \leqslant \|\mathbf{u}\| + \|\mathbf{v}\|$.

A general class of norms, the L_p norms, that satisfy these conditions is given by

$$L_p = \left\{ \sum_i |v_i|^p \right\}^{1/p}$$

where the v_i are the components of $\mathbf{v}$. Note that the Euclidean length is an L_2 norm. The L_1 norm is the sum of the absolute values of the components, and is sometimes called the "taxicab norm." (Think of taking a trip from one point to another by taxi in a city with a square grid of streets. The L_2 norm might be called the "as the crow flies" norm by analogy.) If we let p increase without limit, the value of L tends toward the absolute value of the largest component. Thus we define the L_∞ norm as

$$L_\infty = \max_i |v_i|$$

This norm simply uses the largest component as a measure of the size of a vector.

We define a matrix norm in terms of its effect on a vector. A square matrix A multiplied into a vector $\mathbf{x}$ yields another vector $\mathbf{y}$, which may differ from $\mathbf{x}$ in both magnitude and direction (unless A is an identity matrix). Suppose we examine all combinations of **x** and **y** and choose a pair for which the ratio of $\|\mathbf{y}\|/\|\mathbf{x}\|$ is the greatest. This can be used as a norm for the matrix:

$$\|A\| = \max_{\mathbf{x}} \frac{\|A\mathbf{x}\|}{\|\mathbf{x}\|}$$

This norm is called a *subordinate*, or *induced*, norm because it is defined in terms of a vector norm.

From this definition it can be shown that for square matrices A and B,

$$\|AB\| \le \|A\| \cdot \|B\|$$

and that for all matrices A and vectors **x**,

$$\|A\mathbf{x}\| \le \|A\| \cdot \|\mathbf{x}\|$$

We will need these results in order to understand the material in the rest of the chapter, and we now return to the main topic of the chapter, the solution of simultaneous linear equations.

4.3 Gaussian Elimination

One of the oldest and most useful methods of solving a system of linear equations $A\mathbf{x} = \mathbf{b}$ is Gaussian elimination. The method involves the systematic subtraction of multiples of equations from other equations in order to eliminate variables from some of the equations until the system is upper triangular. In an upper triangular system, the matrix of the system has only zeros below the main diagonal. Thus the first equation contains all the variables, the second equation is missing x_1, the third equation is missing x_1 and x_2, . . . , and the last (nth) equation contains only x_n. At this point, the last equation can be solved for x_n. Then the next-to-last equation can be solved for x_{n-1} after the value of x_n has been substituted. This process of back-solving can be continued until the first equation is solved for x_1, with the values of all the other x's substituted appropriately. The initial triangularization process is sometimes known as *decomposition*.

Example 4.3

$$\begin{aligned} x + y + z &= 6 & \text{(4.1a)} \\ 2x - y + 3z &= 9 & \text{(4.1b)} \\ x + 2y - z &= 2 & \text{(4.1c)} \end{aligned}$$

Multiply (4.1a) by 2 and subtract it from (4.1b):

$$\begin{aligned} x + y + z &= 6 & \text{(4.2a)} \\ 0x - 3y + z &= -3 & \text{(4.2b)} \\ x + 2y - z &= 2 & \text{(4.2c)} \end{aligned}$$

Multiply (4.2a) by $+1$ and subtract it from (4.2c):

$$x + y + z = 6 \tag{4.3a}$$
$$-3y + z = -3 \tag{4.3b}$$
$$0x + y - 2z = -4 \tag{4.3c}$$

Multiply (4.3b) by $-\frac{1}{3}$ and subtract it from 4.3c):

$$x + y + z = 6 \tag{4.4a}$$
$$-3y + z = -3 \tag{4.4b}$$
$$-\frac{5}{3}z = -5 \tag{4.4c}$$

We now solve (4.4c) to obtain $z = 3$. Next we solve (4.4b) for y:

$$y = \frac{-(-3 - z)}{3} = \frac{-(-3 - 3)}{3} = 2$$

Finally, we solve (4.4a) for x:

$$x = 6 - y - z = 6 - 2 - 3 = 1 \quad \square$$

These two sets of operations, Gaussian elimination and back-solving, are relatively easily programmed. We use the concept of an augmented matrix to store the matrix of coefficients and the vector **b**:

$$\begin{bmatrix} 1 & 1 & 1 & 6 \\ 2 & -1 & 3 & 9 \\ 1 & 2 & -1 & 2 \end{bmatrix}$$

We perform the set of elimination operations using 4-digit decimal arithmetic:

$$\begin{bmatrix} 1.000 & 1.000 & 1.000 & 6.000 \\ 2.000 & -1.000 & 3.000 & 9.000 \\ 1.000 & 2.000 & -1.000 & 2.000 \end{bmatrix}$$

Step 1:

$$\begin{bmatrix} 1.000 & 1.000 & 1.000 & 6.000 \\ 0.000 & -3.000 & 1.000 & -3.000 \\ 1.000 & 2.000 & -1.000 & 2.000 \end{bmatrix}$$

Step 2:

$$\begin{bmatrix} 1.000 & 1.000 & 1.000 & 6.000 \\ 0.000 & -3.000 & 1.000 & -3.000 \\ 0.000 & 1.000 & -2.000 & -4.000 \end{bmatrix}$$

Step 3:

$$\begin{bmatrix} 1.000 & 1.000 & 1.000 & 6.000 \\ 0.000 & -3.000 & 1.000 & -3.000 \\ 0.000 & 0.000 & -1.667 & -5.000 \end{bmatrix}$$

Back-solving gives:

$$z = \frac{-5.000}{-1.667} = 2.999$$

$$y = \frac{-(-3.000 - 2.999)}{3.000} = 2.000$$

$$x = 6.000 - 2.000 - 2.999 = 1.001$$

Here we see the effects of round-off error as it occurs in a computer solution.

4.4 Gauss-Jordan Method and Operations Count

Another method, known as Gauss-Jordan reduction, involves a *complete* elimination. In a complete elimination, zeros are produced for all off-diagonal elements, both above and below the main diagonal. In some algorithms, the pivot, or the element used in producing the zeros, is changed to 1 by dividing the row of the matrix through by the original value of the pivot element. Thus matrix A turns into matrix I by this operation. If matrix A is augmented (see Figure 4.2) by matrix I and a Gauss-Jordan reduction is performed on the augmented matrix, then matrix I will turn into matrix A^{-1}. This is a standard way to produce a matrix inverse.

There is a temptation to believe that an effective way to solve a system $Ax = \mathbf{b}$ is to invert A in order to produce $\mathbf{x} = A^{-1}\mathbf{b}$. On the surface this appears to be attractive, particularly when there are several sets of equations to be solved

FIGURE 4.2

$$\left[\begin{array}{cccc|cccc} a_{11} & a_{12} & \cdots & a_{1n} & 1 & 0 & 0 & \cdots & 0 \\ a_{21} & a_{22} & \cdots & a_{2n} & 0 & 1 & 0 & \cdots & 0 \\ \cdot & \cdot & & \cdot & & \cdot & \cdot & \cdot & \\ \cdot & \cdot & & \cdot & & \cdot & \cdot & \cdot & \\ \cdot & \cdot & & \cdot & & \cdot & \cdot & \cdot & \\ a_{n1} & a_{n2} & \cdots & a_{nn} & 0 & 0 & 0 & \cdots & 1 \end{array}\right]$$

(a) Matrix A augmented by unit matrix, I_n.

$$\left[\begin{array}{cccc|cccc} 1 & 0 & \cdots & & 0 & c_{11} & c_{12} & \cdots & c_{1n} \\ 0 & 1 & \cdots & & 0 & c_{21} & c_{22} & \cdots & c_{2n} \\ \cdot & \cdot & & & \cdot & \cdot & & & \cdot \\ \cdot & \cdot & & & \cdot & \cdot & & & \cdot \\ \cdot & \cdot & & & \cdot & \cdot & & & \cdot \\ 0 & 0 & \cdots & & 1 & c_{n1} & c_{n2} & \cdots & c_{nn} \end{array}\right]$$

(b) After Gauss-Jordan reduction, the unit matrix is replaced by $A^{-1} = (c_{ij})$.

all involving the same matrix A but different vectors $\mathbf{b}$. (This occurs quite often, for example, in structures problems, where A characterizes the structure and vector $\mathbf{b}$ represents a set of loads on the structure producing different stresses in the members. We are then interested in solving for the different stress vectors $\mathbf{x}$.) As we shall see, the triangularization (Gaussian elimination) involves significantly fewer arithmetic operations than complete (Gauss-Jordan) elimination. In fact, if we do need the inverse of A for some reason (other than the solution of $A\mathbf{x} = \mathbf{b}$), it is much more efficient to solve n equations of the form

$$A\mathbf{x} = \mathbf{c}$$

where $\mathbf{c}$ is one of the columns of the identity matrix I. Good Gaussian elimination programs preserve the triangularized form of A and allow back-solving with any number of different right-hand-side vectors.

We next look at the operation count for Gaussian elimination. Traditionally, we look only at the number of multiplications and divisions because these tended in the past to dominate the time required for the reduction to triangular form. Although this is less and less the case now, it is still a good rough measure because the number of other arithmetic operations tends to increase in proportion to the number of these operations. In the first equation, $a_{12}, \ldots , a_{1n}$ must each be multiplied by the quantity a_{21}/a_{11}, which amounts to $n - 1$ multiplications preceded by one division, for a total of n such operations. This is done for each of the $n - 1$ equations below the first equation, so that we have a total of $n(n - 1)$ multiplication or division operations to put zeros in the $n - 1$ locations in the first column. For the second equation we have only $n - 2$ equations below that equation and $n - 2$ elements to the right of the pivot element. If we add the division involved in producing the multiplier, we have $(n - 1)(n - 2)$ multiplication and division operations. For each succeeding operation, the factors in the product are reduced by one: $(n - 2)(n - 3)$, $(n - 3)(n - 4)$, $\ldots$, $(2)(1)$. We can add these terms to get an expression $n^3/3 - n/3$ as the total number of multiplications and divisions. Proving this expression is left as an exercise.

For the back-solving, we find that to solve for x_n we must do one division:

$$x_n = \frac{b'_n}{a'_{nn}}$$

To find x_{n-1} we must do one division and one multiplication:

$$x_{n-1} = \frac{b'_{n-1} - a'_{n-1, n} \cdot x_n}{a'_{n-1, n-1}}$$

As we go back up the set of equations, at each step we add one more term requiring a multiplication, so that the total number of multiplication or division steps is

$$1 + 2 + 3 + 4 + \cdots + n = \frac{n^2}{2} + \frac{n}{2}$$

For the first equation, from which we find x_1, there will be $n - 1$ $a_{ij} \cdot x_j$ products and 1 division.

If we repeat this computation for the Gauss-Jordan reduction, we see that the first zero in the first column requires (as before) n multiplication and division operations. Thus putting the $n - 1$ zeros in the first column requires $n(n - 1)$ multiplications and divisions. Now, however, the second column of zeros requires $(n - 1)(n - 1)$ operations because we must put a zero above the main diagonal. In fact, we must put $n - 1$ zeros in each column, so that the sum of terms is

$$n(n - 1) + (n - 1)(n - 1) + (n - 2)(n - 1) + \cdots + 2(n - 1) + (n - 1)$$
$$= \frac{(n - 1)(n + 1)n}{2} = \frac{n^3}{2} - \frac{n}{2}$$

Here, of course, there is no back-solving to be done.

For a system as small as 10 equations we would have about 330 operations for Gaussian elimination plus 55 for back-solving, or a total of 385 multiplication and division steps. For Gauss-Jordan we would have $500 - 5$, or 495, operations. For large systems the multiplication and division work for Gaussian triangularization with back-solving tends toward a factor of two-thirds of the work done in Gauss-Jordan complete reduction. This is not to be ignored for large problems.

4.5 Sources of Trouble

If we naively apply Gaussian elimination to some matrices, we will experience disaster. A simple example will suffice to show the source of the trouble. The following matrix, A, is nonsingular.

$$A = \begin{bmatrix} 1 & 2 & 4 \\ 1 & 2 & -5 \\ 5 & 1 & 10 \end{bmatrix}$$

After the first two steps we have

$$A' = \begin{bmatrix} 1 & 2 & 4 \\ 0 & 0 & -9 \\ 0 & -9 & -10 \end{bmatrix}$$

If we attempt to continue in the simple way as shown before, we will have a zero pivot, an attempted division by zero, and failure. It is clear that we could have avoided this state of affairs by exchanging the last two equations, or the last two rows in the matrix. We must develop a method for avoiding the necessity of having the equations in a particular order because, in general, this is difficult to accomplish by inspection in very large systems.

Zeros in pivot positions can occur in different ways. One situation might be called the occurrence of an accidental zero, such as in the example, where the system does have a perfectly good solution. This case can be cared for easily by the pivoting strategy to be described. More difficult problems arise when the matrix is either *ill-conditioned* (*nearly* singular) or singular. In the latter case, most often near the end, a very small pivot may be found that should be zero but is not because of round-off error. In the former case, a zero pivot may be encountered that should be a small, nonzero number were it not for round-off error. It is sometimes difficult to know whether a matrix is truly singular or whether it is ill-conditioned. Good software will provide a method of identifying ill-conditioned linear systems, but it cannot be expected to discriminate reliably between the truly singular and the nearly singular cases.

In order to avoid an accidental zero pivot and also to improve the accuracy of the result by avoiding very small pivots that may have lost accuracy through subtraction, we use what is called a *pivot strategy*. In partial pivoting we start by scanning the first column to find the element of largest absolute value,

$$\text{Pivot} = \max_{i} |a_{i1}|$$

We then exchange the equation in which it is found with the first equation. After the first column of zeros has been found, we scan the second column below the first row to find

$$\text{Pivot} = \max_{i} |a_{i2}| \qquad i \neq 1$$

The equation in which the pivot is found then becomes the second equation in the set. This process is continued until the system has been made upper triangular. If the matrix is nonsingular, no zero pivot will ever be found unless the matrix is so poorly conditioned that round-off error converts a very small pivot to zero.

Complete pivoting is a slower and more elaborate process. In this method the first pivot is selected by scanning the entire matrix:

$$\text{Pivot} = \max_{i,j} |a_{ij}|$$

This number is used as the first pivot. After the first set of zeros has been created in the jth column, the matrix is again scanned over all rows and columns except for the row and column of the first pivot. Obviously this method requires much more work in establishing the pivots to be used. Furthermore, it is much more complicated to keep track of the proper ordering of rows and columns. It does not usually give significantly better results than partial pivoting. Therefore, it is used less often.

4.6 Equilibration

The purpose of partial pivoting is to ensure the choice of large rather than small pivots in an attempt to avoid accidents and to improve the accuracy of the solution. Through circumstances of the scaling of the equations, the largest prospective pivot in a column may be the smallest number in its row. If the roles of row and column were reversed, that element would not be a candidate. This situation can be prevented by equilibration (by row), which is the process of dividing each equation through by the magnitude of its largest coefficient to make the L_∞ norm equal to 1. (Using the L_2 norm would also be satisfactory, but it requires more computation.) Generally, equilibration seems to be beneficial. However, the explicit process tends to increase the round-off error because of the increased number of arithmetic operations. The round-off problem can be avoided in binary machines by using as the divisor the smallest power of 2 that is equal to or greater than the magnitude of the largest coefficient in the matrix row. Such a practice means that only exponent manipulations occur. In the *implicit equilibration* method, the pivot candidates in each column are divided by the largest element in the candidate's row. The largest such quotient is chosen, but the actual pivoting is done by the undivided element. This avoids the extra round-off error created by simple scaling of the equations. Examples can be contrived to show that under certain conditions an equilibrated system may behave less well under solution than the original system. Fortunately, such problems are seldom encountered in run-of-the-mill applications.

4.7 Condition

In this section we look at the concept of the condition of a linear system in more detail. We shall decide what it means mathematically when we say that a system is ill-conditioned. We also consider how we can determine if a given system is ill-conditioned and thus is likely to give bad results upon solution.

The idea of condition can be understood in a geometrical context, which gives us some insight into why a poorly conditioned system often has substantial errors in its solution. Figure 4.3 shows the graphs of two straight lines whose equations are $x + y = 5$ and $x - y = -1$. The problem of solving the system

$$\begin{bmatrix} 1 & 1 \\ 1 & -1 \end{bmatrix} \begin{bmatrix} x \\ y \end{bmatrix} = \begin{bmatrix} 5 \\ -1 \end{bmatrix}$$

is equivalent to finding the point of intersection of the graphs of the equations. Figure 4.4 shows the graphs of $x + y = 5$ and $12x + 10y = 54$. We see that the lines are almost parallel. If we now realize that the operations of arithmetic produce slight perturbations in the coefficients of the equations, thus slightly perturbing the positions of the lines in the graphs, we recognize that in the case of Figure 4.3 this leads to a small uncertainty in the position of the intersection (whose coordinates are the solution of the equations). In the case of Figure 4.4, however, the region of uncertainty is much larger. The more nearly the lines are to being parallel, the less certainty we have about the true solution. For example, the pair of equations

$$x + y = 1$$
$$x + 1.01y = 2$$

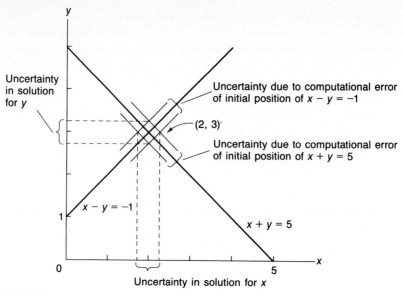

FIGURE 4.3

has the exact solution $(-99, +100)$. The system

$$x + y = 1$$
$$1.01x + y = 2$$

has the exact solution $(+100, -99)$. Not only does the solution change greatly with very small changes in the coefficients, but the computed solutions show appreciable errors. Suppose we assume that the coefficients are exact and solve the two sets of equations with the programs shown later in this chapter. We find that the computed solutions are

$$(-99.000000016, +100.00000002)$$

and

$$(+100.00000000, -99.000000003)$$

This example illustrates the effects of ill-conditioning: (1) the solution changes drastically for small changes in the coefficients; and (2) the computed solution shows great magnification of the effects of round-off error, which produces errors in the computed solution sometimes many orders of magnitude greater than the machine epsilon. Well-conditioned systems are relatively stable against perturbations of the coefficients. The computed solutions tend to be in error by amounts that are seldom more than a small multiple of the machine epsilon.

In the well-conditioned system (Figure 4.3), the computed solution, using the same computer and program as earlier, is $(2.00000000000, 3.00000000000)$. If

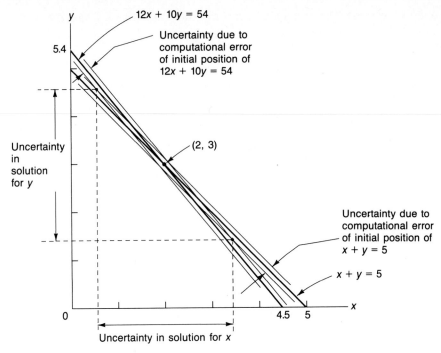

FIGURE 4.4

we change the coefficient of x to 1.01 in the equation $x + y = 5$, the solution becomes (1.9900497512, 2.9900497512). If, instead, we change the y coefficient to 1.01, the solution becomes (1.9850746269, 2.9850746269). Note that the computer solution of the original problem is correct to the number of places shown (the machine epsilon in this case is $2^{-39} = 1.82E-12$) and that perturbations in the coefficients of A on the order of 1% produce perturbations in the solution of about the same relative size.

Ill-conditioning is in a sense a measure of the "closeness" of the coefficient matrix to a singular matrix. We know that the matrix

$$\begin{bmatrix} 1 & 1 \\ 1 & 1 \end{bmatrix}$$

is a singular matrix, so that we are not surprised to find that the matrices

$$\begin{bmatrix} 1 & 1 \\ 1.01 & 1 \end{bmatrix} \quad \text{and} \quad \begin{bmatrix} 1 & 1 \\ 1 & 1.01 \end{bmatrix}$$

are close to being singular. However, the matrix

$$\begin{bmatrix} 1 & -1 \\ 1 & 1 \end{bmatrix}$$

is not so close. We might be tempted to use the determinant of A as a measure of closeness because we know that

$$\det \begin{bmatrix} 1 & 1 \\ 1 & 1 \end{bmatrix} = 0, \qquad \det \begin{bmatrix} 1 & 1 \\ 1 & 1.01 \end{bmatrix} = 0.01$$

and

$$\det \begin{bmatrix} 1 & -1 \\ 1 & 1 \end{bmatrix} = 2$$

Unfortunately the matrix

$$\begin{bmatrix} 100 & 100 \\ 100 & 101 \end{bmatrix}$$

is just as poorly conditioned as

$$\begin{bmatrix} 1 & 1 \\ 1 & 1.01 \end{bmatrix}$$

but its determinant is 100.

In order to investigate condition we start with $Ax = b$ and see what happens if we perturb the vector $\mathbf{b}$ slightly to produce a new vector $\mathbf{b} + \delta\mathbf{b}$, where $\delta\mathbf{b}$ is small in the sense that $\|\delta\mathbf{b}\|$ is small relative to $\|\mathbf{b}\|$. Then

$$A(\mathbf{x} + \delta\mathbf{x}) = \mathbf{b} + \delta\mathbf{b}$$

where $\delta\mathbf{x}$ is the perturbation of $\mathbf{x}$. Since $Ax = \mathbf{b}$, we see that

$$A\delta\mathbf{x} = \delta\mathbf{b}$$

or

$$\delta\mathbf{x} = A^{-1}\delta\mathbf{b}$$

Now $\|A\mathbf{x}\| = \|\mathbf{b}\|$, and $\|\mathbf{b}\| \leq \|A\| \cdot \|\mathbf{x}\|$. Also, we can write $\|\delta\mathbf{x}\| \leq \|A^{-1}\| \cdot \|\delta\mathbf{b}\|$, which enables us to divide the inequalities to find

$$\frac{\|\delta\mathbf{x}\|}{\|A\| \cdot \|\mathbf{x}\|} \leq \frac{\|A^{-1}\| \cdot \|\delta\mathbf{b}\|}{\|\mathbf{b}\|}$$

or

$$\frac{\|\delta\mathbf{x}\|}{\|\mathbf{x}\|} \leq \|A\| \cdot \|A^{-1}\| \frac{\|\delta\mathbf{b}\|}{\|\mathbf{b}\|}$$

This tells us that the relative change in the solution vector (using norms for measurement) can be as great as $\|A\| \cdot \|A^{-1}\|$ times as much as the change in the vector of constants. We call the quantity $\|A\| \cdot \|A^{-1}\|$ the condition number of the matrix A and designate it as COND(A) or $K(A)$. The value of $K(A)$ is bounded below by $+1$. As the condition of the matrix gets worse, the value of $K(A)$ increases without limit and is undefined for a singular matrix.

Next we look at the problem of what happens to the solution when we perturb the elements of A. In this case we have

$$(A + \delta A)(x + \delta x) = b$$

which is the same as

$$\delta Ax + A\delta x + \delta A\delta x = 0$$

because $Ax = b$. From this we can find

$$\delta x = -A^{-1}\,\delta A(x + \delta x)$$

or

$$\|\delta x\| \leqslant \|A^{-1}\| \cdot \|\delta A\| \cdot \|x + \delta x\|$$

This can be divided through by $\|x + \delta x\|$ to find

$$\frac{\|\delta x\|}{\|x + \delta x\|} \leqslant \|A^{-1}\| \cdot \|\delta A\|$$

or

$$\frac{\|\delta x\|}{\|x + \delta x\|} \leqslant \|A\| \cdot \|A^{-1}\| \cdot \frac{\|\delta A\|}{\|A\|}$$

Again we find that the relative change in the solution is bounded by the condition number times the relative change in the matrix, as measured by the norms.

The condition number of

$$\begin{bmatrix} 1 & 1 \\ 1.01 & 1 \end{bmatrix}$$

can be found by finding first the inverse

$$\begin{bmatrix} -100 & 100 \\ 101 & -100 \end{bmatrix}$$

and then taking L_1 norms: $\|A\| = 2.01$ and $\|A^{-1}\| = 201$. Then we find that $K(A) = 404 = 4.04 \times 10^2$. A rule of thumb is that the exponent on the power of 10 when $K(A)$ is expressed in power-of-ten form is a rough indicator of the number of decimals of lost accuracy in the solution, given that all coefficients are considered exact. Note that this is consistent with what we found in solving

$$\begin{bmatrix} 1 & 1 \\ 1 & 1.01 \end{bmatrix}\begin{bmatrix} x \\ y \end{bmatrix} = \begin{bmatrix} 1 \\ 2 \end{bmatrix}$$

where we found about two decimal places of error in the value of x.

Finding $K(A)$ by using $\|A\| \cdot \|A^{-1}\|$, which is found by first inverting A, is not a practical method. For one thing, the inversion requires more computation than the triangularization of A. For another, the inverse is likely to be inaccurate when A is poorly conditioned, so that the computation of $K(A)$ is uncertain. For these reasons a different method of estimating $K(A)$ is used. This method is relatively fast; although it is also inaccurate, the time required for making the estimate is much less than that of the inversion of A. With this method the estimate of the condition number generally is bounded above by the true (theoretical) condition number and below by the true condition number divided by the order of the matrix. For most purposes this is good enough, provided that the user is aware of the nature of the estimate.

4.8 Estimating $K(A)$

The definition of the condition number of a matrix A is that $K(A) = \|A\| \cdot \|A^{-1}\|$, where $\| \cdot \|$ is some norm. Computing A^{-1} requires more work than we wish to do. An easy way out involves a knowledge of the behavior of norms.

Suppose $\mathbf{y}$ and $\mathbf{z}$ are vectors such that $\mathbf{y} = A^{-1}\mathbf{z}$. But we know that $\|\mathbf{y}\| \leqslant \|A^{-1}\| \cdot \|\mathbf{z}\|$, so that $\|A^{-1}\| \geqslant \|\mathbf{y}\|/\|\mathbf{z}\|$. Now suppose further that we choose a number of random vectors $\mathbf{z}_i$ and solve $A\mathbf{y}_i = \mathbf{z}_i$ for the corresponding $\mathbf{y}_i$. Then we choose the $(\mathbf{y}_i, \mathbf{z}_i)$ pair that maximizes $\|\mathbf{y}_i\|/\|\mathbf{z}_i\|$ and let this be the estimate of $\|A^{-1}\|$. By triangularizing A, we can solve any problem of the form $A\mathbf{y} = \mathbf{z}$ with order of n^2 operations. We could try half a dozen random vectors $\mathbf{z}$ to find the maximum $\|\mathbf{y}\|/\|\mathbf{z}\|$ for only a small additional amount of work. It is estimated that the average $\|A^{-1}\|$ estimated on the basis of one random vector $\mathbf{z}$ is about half of the true value of $\|A^{-1}\|$. Thus an estimate based on half a dozen random $\mathbf{z}$'s ought to be very good.

A more sophisticated approach has been used by Cline, Moler, Stewart, and Wilkinson [1979]. This approach is used in the LINPACK programs as well as in GAUSS in this book. The procedure uses a vector of $+1$ or -1 values in the components of $\mathbf{z}$ chosen in such a way as to try to maximize $\|\mathbf{y}\|$.

4.9 Iterative Improvement

Iterative improvement is a technique for improving the solution of a linear system. Suppose $\mathbf{x}'$ is the solution computed for the system $A\mathbf{x} = \mathbf{b}$. Then if we let $\mathbf{r} = \mathbf{b} - A\mathbf{x}'$ be the residual, or the amount by which $\mathbf{x}'$ fails to satisfy the set of equations, we let $\delta\mathbf{x} = \mathbf{x} - \mathbf{x}'$ be the correction to $\mathbf{x}'$. Now we know that $A(\mathbf{x} - \mathbf{x}') = \mathbf{b} - A\mathbf{x}' = \mathbf{r} = A\delta\mathbf{x}$, so we might try solving the equation $A\delta\mathbf{x} = \mathbf{r}$ for $\delta\mathbf{x}$. Obviously, if A has been triangularized, the solution will be very cheap. However, unless the precision of the computation is increased (which usually means at least double precision in the calculation of $\mathbf{r}$), the error in the calculation of $\delta\mathbf{x}$ is on the order of $\mathbf{r}$ itself. This makes the process rather costly in time and to some extent in storage. The process can be repeated as many times as necessary until the solutions $\mathbf{x}$ no longer change from one cycle to the next.

Iterative improvement is often included as an option in software systems, but its use is expensive.

4.10   Procedures GAUSS and SOLVE

GAUSS is a PASCAL procedure that uses Gaussian elimination to convert a matrix to upper triangular form. It uses partial pivoting but it does not use equilibration. It also incorporates the estimation of the condition number, using the technique described in the previous section.

The call to GAUSS is

```
GAUSS (N, A, COND, IPVT) ;
```

In the calling program A must be declared in a TYPE statement as follows:

```
TYPE        MATRIX = ARRAY[1..SIZE, 1..SIZE] OF REAL;
```

where SIZE is an appropriate integer for the system. Then A is declared in a VAR statement:

```
VAR        A:        MATRIX;
```

IPVT must also be declared:

```
TYPE       INT_VECTOR = ARRAY[1..SIZE] OF INTEGER;
VAR             IPVT:    INT_VECTOR:
```

Upon return A contains the triangularized version in the upper triangle. Since the lower triangle would otherwise be all zeros, it is possible to store information there. This allows for storage of all the pivots needed to work on the **b** vector in the SOLVE phase. The integer vector IPVT stores the pivoting information from the Gaussian elimination phase. This is also needed by SOLVE. COND contains, upon return, an estimate of the condition number. If the procedure encounters machine zeros for all possible pivots at any point, the procedure will return with 1.0E + 32 as a signal that there seems to be a singularity.

It is possible to show that the product of the diagonal elements of a Gaussian-reduced matrix is the value of the determinant of the matrix. To obtain the determinant, use the statements:

```
DET := 1.0;
FOR I := 1 TO N DO  DET := DET*A[I,I];
DET := DET*IPVT[N];
```

IPVT [N] corrects the sign of the product in order to make the sign correspond to the original ordering of rows in the matrix before pivoting caused any exchanges.

Procedure SOLVE accepts the output of GAUSS as input and does the back-solving necessary to produce a solution. The call to SOLVE is as follows:

```
SOLVE (N, A, B, IPVT) ;
```

N is the order of A, and A is the name of the matrix that has been triangularized in GAUSS. B is the right-hand-side vector of constants, and IPVT is the vector of integers containing the pivot information necessary to match up the rows of B with the rows of A. The solution to $Ax = \mathbf{b}$ will be returned in the vector B. SOLVE should never be called if COND is very large and, in particular, if COND is equal to the flag number 1.0E + 32.

Example 4.4

Find the solution to the following system of linear equations:

$$2x - 3y + 10z = 58$$
$$x + y + 3z = 14$$
$$3x + 10y + z = -12$$

Use GAUSS and SOLVE. The program to solve this system is shown in Figure 4.5. (In most cases we would not set the values of the coefficients in the program but would provide input routines to allow entry of coefficients from a keyboard, from disk storage, or even from punched cards.) Figure 4.6 shows the program output. The exact solution is (1, −2, 5). The error in the value of x is consistent with the size of the condition number. That is, we have lost about two digits of accuracy in one solution component. □

```
Program Gauss_Demonstration;

Type
      Matrix = Array[1..10,1..10] of Real;
      Vector = Array[1..10] of Real;
   Int_Vector = Array[1..10] of Integer;

Var
            A: Matrix;
            B: Vector;
         Ipvt: Int_Vector;
            N: Integer;
            I: Integer;
         Cond: Real;
          Det: Real;

{$I B:Solve.pas}
{$I B:Gauss.pas}

Begin

      A[1,1] := 2.0;    A[1,2] := -3.0;   A[1,3] := 10.0;
```

```
A[2,1] := 1.0;    A[2,2] := 1.0;    A[2,3] := 3.0;
A[3,1] := 3.0;    A[3,2] := 10.0;   A[3,3] := 1.0;

B[1] := 58.0;     B[2] := 14.0;     B[3] := -12.0;

N := 3;
Gauss(N,A,Cond,Ipvt);
If Cond > 1.0E+20 then
   Begin
      Writeln('Condition number exceeds 1.0 E+20; Halt.');
      Halt
   End;
Solve(N,A,B,Ipvt);

Writeln; Writeln('Solution is: '); Writeln;
For I := 1 to N do Writeln(B[I]);
Writeln;
Writeln('Condition number = ',Cond); Writeln;
Det := 1.0;
For I := 1 to N do
   Det := Det*A[I,I];
Det := Det*Ipvt[N];
Writeln('Determinant of matrix is:   ',Det)

End.
```

FIGURE 4.5

```
Solution is:

   9.9999999989E-01
  -2.0000000000E+00
   5.0000000000E+00

Condition number:     1.8121476940E+02

Determinant of matrix is:   -1.2000000000E+01
```

FIGURE 4.6

4.11 Procedures TRIDIAG and PENTDIAG

Occasionally systems of linear equations arise that have a special form for the matrix A. The first of these is a form in which all elements of the matrix are zero except the elements lying on, immediately above, or immediately below the main diagonal. In other words, $A[I, J] = 0$ unless $|I - J| < 2$. Such a system is called *tridiagonal*. Another case of interest is one for which $A[I, J] = 0$

unless $|I - J| < 3$. These systems are particularly attractive because the elements can be stored in 3N or 5N locations. The solution time is proportional to N rather than to N^3.

TRIDIAG and PENTDIAG are PASCAL procedures for solving tridiagonal and five-diagonal systems of linear equations. The call to TRIDIAG is

TRIDIAG (N, A, B, C, D) ;

where N is the order of the system, A is a vector containing the subdiagonal elements, D is a vector of the main diagonal elements, C is a vector of the superdiagonal elements, and B is a vector of the right-hand-side constant elements.

The elements are numbered from their beginnings, which means that the matrix looks like the following in the upper left corner:

$$
\begin{bmatrix}
D[1] & C[1] & 0 & 0 & \cdots \\
A[1] & D[2] & C[2] & 0 & \cdots \\
0 & A[2] & D[3] & C[3] & \cdots \\
0 & 0 & A[3] & D[4] & \cdots \\
0 & 0 & 0 & A[4] & \cdots \\
\vdots & \vdots & \vdots & \vdots & \vdots
\end{bmatrix}
$$

Each of the vectors A, B, C, and D must be declared as TYPE VECTOR by the following kinds of statements:

TYPE   VECTOR   =   ARRAY [1..SIZE] OF REAL;

and then

VAR   A, B, C, D: VECTOR;

where SIZE is the expected maximum vector length. B contains the vector of answers upon return.

Example 4.5

Solve the following tridiagonal system of linear equations:

$$
\begin{aligned}
5x + y \quad\quad\quad &= 7 \\
x + 5y + 2z \quad\quad &= 9 \\
y + 5z + 3w &= 6 \\
z + 5w &= 14
\end{aligned}
$$

Use TRIDIAG. Note that the system is diagonally dominant, so that we can use TRIDIAG with no fear that an accidental zero will occur as a pivot or that the solution will be poor because of ill-conditioning. Figure 4.7 shows the program that calls TRIDIAG to do the solution. The solution is shown in Figure 4.8. The solution is correct to the number of decimal places shown.

```
Program Tridiag_Demonstration;

Type
      Vector = Array[1..10] of Real;

Var
      A,B,C,D: Vector;
            I: Integer;
            N: Integer;

{$I B:Tridiag.pas}

Begin

      D[1] := 5.0; D[2] := 5.0; D[3] := 5.0; D[4] := 5.0;
      A[1] := 1.0; A[2] := 1.0; A[3] := 1.0;
      C[1] := 1.0; C[2] := 2.0; C[3] := 3.0;
      B[1] := 7.0; B[2] := 9.0; B[3] := 6.0; B[4] := 14.0;

      N := 4;

      Tridiag(N,A,B,C,D);

      Writeln; Writeln('Solution is: '); Writeln;
      For I := 1 to N do
        Writeln(B[I])

End.
```
FIGURE 4.7

```
Solution is:

   1.0000000000E+00
   2.0000000000E+00
  -1.0000000000E+00
   3.0000000000E+00
```
FIGURE 4.8

□

PENTDIAG is simply an expansion of TRIDIAG to a five-diagonal banded linear equation system. The call to PENTDIAG is

```
PENTDIAG(N,A,B,C,D,E,F);
```

All features of the call are the same as in TRIDIAG except that there is a sub-subdiagonal provided by vector E and a super-superdiagonal provided by vector F. As in TRIDIAG, the answers come back in vector B. The TYPE declarations and VAR declarations are exactly the same as in TRIDIAG, except that they must include E and F. The arrangement of the matrix is

$$
\begin{bmatrix}
D[1] & C[1] & F[1] & 0 & 0 & \cdot & \cdot & \cdot \\
A[1] & D[2] & C[2] & F[2] & 0 & \cdot & \cdot & \cdot \\
E[1] & A[2] & D[3] & C[3] & F[3] & \cdot & \cdot & \cdot \\
0 & E[2] & A[3] & D[4] & C[4] & \cdot & \cdot & \cdot \\
0 & 0 & E[3] & A[4] & D[5] & \cdot & \cdot & \cdot \\
0 & 0 & 0 & E[4] & A[5] & \cdot & \cdot & \cdot \\
0 & 0 & 0 & 0 & E[5] & \cdot & \cdot & \cdot \\
\cdot & \cdot & \cdot & \cdot & \cdot & \cdot & & \\
\cdot & \cdot & \cdot & \cdot & \cdot & & \cdot & \\
\cdot & \cdot & \cdot & \cdot & \cdot & & & \cdot
\end{bmatrix}
$$

In both algorithms Gaussian elimination and back-solving are used, but in both cases the computing time is linear in the size of the system. No pivoting or equilibration is provided in either algorithm. If the systems to be solved are diagonally dominant—that is, if the absolute value of the diagonal element is greater than the sum of the absolute values of the off-diagonal elements in all rows (or columns)—then the equations are well-conditioned. Furthermore accidental zero pivots cannot occur.

4.12   Some IMSL Subroutines

Some PASCAL systems allow the user to call FORTRAN subprograms. If you wish to use IMSL subroutines, for example, you will need to check with a system programmer to find out how it is done on your system.

The IMSL library contains a variety of linear-equation solvers and matrix inverters for general and special cases. The subroutine names beginning with LEQ are for systems of linear equations. The other families have, for example, names beginning with LINV for matrix inversion. In the LEQ series there are programs for the general case of linear equations with real coefficients and no special properties of the matrix and for various special cases in which special properties of the matrices can be exploited. Each is available in two versions: *space economizer* and *high accuracy*. The latter incorporates iterative improvement if that is needed.

Full-storage mode is provided and needs no explanation; this is the normal array storage of FORTRAN (columns of an array are stored end to end). Symmetric storage mode takes advantage of the fact that $a_{ij} = a_{ji}$ for all i and j, so that only the lower triangle, including the main diagonal, is stored. The lower triangle is sliced up by rows, which are stored end to end. A 3×3 symmetric matrix is stored in this order:

A(1, 1), A(2, 1), A(2, 2), A(3, 1), A(3, 2), A(3, 3)

The reader is referred to the IMSL manual for the descriptions of banded and banded-symmetric storage modes. Suffice it to say here that the various sub-, super-, and main diagonals are stored in columns of a matrix.

The subroutine LEQT1F is the space-economizer version of the general real-matrix equation solver. LEQT2F is the high-accuracy version, in which the solution is tested as described later; an iterative improvement scheme is called if necessary. The call to LEQT1F is

        CALL LEQT1F (A, N, NEQ, K, B, ITEST, WORK, IER)

A is the matrix, N is the number of solutions desired (number of right-hand vectors stored in B), NEQ is the number of equations (number of actual rows filled in A), K is the number of rows of A in its DIMENSION statement (so that the program will know how to find the beginning of each column in the matrix A), B contains right-hand-side vector(s), WORK is a user-defined work area that must be dimensioned according to the IMSL manual rules, and IER is an error flag. ITEST is a user-supplied integer that specifies the number of digits of accuracy of the constants in A and B. The program checks whether the solution vector is accurate for a linear-equation problem with elements of A and B perturbed by no more than 10^{-N}. This test is used in the case of LEQT2F to initiate iterative improvement if necessary. You should read the IMSL manual carefully and try to understand it before you make use of these subroutines.

4.13 LINPACK Programs

LINPACK programs SGECO and SGEFA are programs in FORTRAN that correspond to GAUSS. They triangularize a matrix by Gaussian elimination. SGECO estimates the reciprocal of the condition number of a matrix, but SGEFA does not. For small matrices, say of order 10, SGECO takes about twice as much time as SGEFA. However, for large matrices, say of the order of 100, SGECO takes only about 10% more time.

After the matrix has been triangularized (and the condition number tested), the subroutine SGESL can be called to back-solve for either

$$Ax = b$$

or

$$A^T x = b$$

A fourth routine, SGEDI, computes the inverse and/or the determinant of A after A has been triangularized.

The subroutine calls are as follows.

        CALL SGECO (A, LDA, N, IPVT, RCOND, Z)

where

- A is the array containing the matrix.
- LDA is the number of rows in the dimension statement for A.
- N is the order of the matrix; N ≤ LDA.
- IPVT is a one-dimensional array of integers containing pivot information about A; dimension ≥ N.
- RCOND is the reciprocal of the condition number.
- Z is a one-dimensional REAL work array; dimension ≥ N.

  CALL SGEFA (A, LDA, N, IPVT, INFO)

where

- A, LDA, N, IPVT are as before.
- INFO (on return) is normally zero, but if there is a zero pivot in the (K, K) position of the upper triangular matrix A, then INFO = K.

  CALL SGESL (A, LDA, N, IPVT, B, JOB)

where

- A, LDA, N, IPVT are as before.
- B is the right-hand-side vector of coefficients on the call, which contains the solution vector on return; dimension ≥ N.
- JOB = 0 if $Ax = \mathbf{b}$ is to be solved, and nonzero for the solution of $A^T\mathbf{x} = \mathbf{b}.$

  CALL SGEDI (A, LDA, N, IPVT, DET, WORK, JOB)

where

- A, LDA, N, IPVT are as before.
- WORK is a one-dimensional REAL array; dimension ≥ N.
- JOB is a request for action:

   JOB = 01 gives inverse only

   JOB = 10 gives determinant only

   JOB = 11 gives both

- DET is an array of length 2 such that

  DET (A) = DET (1) *10**DET (2)

where $1.0 \le |DET (1)| < 10.0$ or DET (1) = 0, and where DET (2) is the REAL form of an integer.

A test for singularity upon exit from SGECO is

IF (1.0 + RCOND = 1.0) THEN matrix is singular

4.14 Iterative Methods

Gaussian elimination is an example of a finite method or one in which a solution can be obtained in a finite, predetermined number of steps. In contrast, there are also iterative solution methods, such as those known as the Jacobi and Gauss-Seidel methods. These tend to be used in certain kinds of problems having matrices that are sparse and banded and that have certain properties we will discuss later.

The solution methods are perhaps easiest to explain with a simple example.

Example 4.6

$$5x + y \quad\quad = 6$$
$$x + 5y + z = 7$$
$$y + 5z = 6$$

which has the solution $(1, 1, 1)$. We solve for x from the first equation, for y from the second, and for z from the third:

$$x = \frac{6 - y}{5}$$

$$y = \frac{7 - x - z}{5}$$

$$z = \frac{6 - y}{5}$$

We now assume a starting vector, say $(0, 0, 0)$, and use this in the set of equations to obtain a new estimate $(1.2, 1.4, 1.2)$. If we repeat the process, we get $(0.92, 0.92, 0.92)$; from that, we get $(1.016, 1.032, 1.016)$. It appears that the process is converging on the exact solution $(1, 1, 1)$.

We can make a modification of this method (Jacobi's) to obtain the Gauss-Seidel method. Instead of substituting $(0, 0, 0)$ in all three of the equations to get $(1.2, 1.4, 1.2)$, we substitute into the first equation and then immediately update the solution vector; we get

$$x = 1.2, \quad y = \frac{7 - 1.2 - 0}{5} = 1.16, \quad z = \frac{6 - 1.16}{5} = 0.968$$

$$x = \frac{6 - 1.16}{5} = 0.968, \quad y = 1.0128, \ldots$$

A table of these solutions is shown for three iterations:

Jacobi	Gauss-Seidel
1.2, 1.4, 1.2	1.2, 1.16, 0.968
0.92, 0.92, 0.92	0.968, 1.0128, 0.99744
1.016, 1.032, 1.016	0.99744, 1.001024, 0.9997952

The L_2 norm of the error between the true solution and the Jacobi solution at the end of three steps is 0.03919, or about 2% of the norm of the solution vector. The L_2 norm of the error for the Gauss-Seidel case is 0.00276, or less than 0.2% of the norm of the solution vector. No more arithmetic is required in one method than the other. However, the Gauss-Seidel method has the advantage of needing only one vector in which to store the solution at any time; the Jacobi method requires two. The main advantage of the Gauss-Seidel method over the Jacobi method is, of course, its speed of convergence.   □

The Jacobi method can be written in matrix form as

$$\mathbf{x}_{n+1} = B\mathbf{x}_n + \mathbf{c}$$

where $\mathbf{x}_n$ and $\mathbf{x}_{n+1}$ are successive iterates of the vector $\mathbf{x}$, $\mathbf{c}$ is a vector whose ith element is b_i/a_{ii}, and B is a matrix whose ijth element is

$$b_{ij} = \begin{cases} 0, & i = j \\ \dfrac{-a_{ij}}{a_{ii}}, & \text{otherwise} \end{cases}$$

The Gauss-Seidel method can also be expressed in matrix form as

$$\mathbf{x}_{n+1} = L\mathbf{x}_{n+1} + U\mathbf{x}_n + \mathbf{c}$$

where $\mathbf{x}_n$, $\mathbf{x}_{n+1}$, and $\mathbf{c}$ are as before, L is the lower triangle (elements below the main diagonal) of B, and U is the upper triangle (elements above the main diagonal) of B. Note that in solving for the $(n + 1)$st value of the first variable, all the elements of the top row of L are zero. Thus no values of $\mathbf{x}_{n+1}$ are involved in finding the first variable value.

Because of its superior convergence, Gauss-Seidel is preferred over Jacobi, so we shall talk mainly about it. For both methods it is clear that no main diagonal element of A in the system $A\mathbf{x} = \mathbf{b}$ can vanish. As a practical matter, convergence is guaranteed if diagonal dominance holds for A, which is one or the other of the following:

$$|a_{ii}| > \sum_{j \neq i} |a_{ij}| \qquad \text{for all } i$$

$$|a_{ii}| > \sum_{j \neq i} |a_{ji}| \qquad \text{for all } i$$

The proofs of convergence are beyond the scope of this book. Young and Gregory [1973] is a good reference.

The diagonal dominance criterion is often easy to apply when the system is very sparse. In many practical cases the equations of the system can be inspected to show whether convergence can be expected. Convergence of the Gauss-Seidel method can be speeded significantly by the use of what is called *successive over-relaxation*. In this procedure one anticipates that if one of the variables changes by an amount q during an iteration cycle, then it is likely to change again and in the same direction in the next cycle. Therefore, at each cycle, the variable is not changed by q, the computed amount of change (new value minus the old value, by the equation) but it is instead changed by $q\omega$, where ω is the successive overrelaxation factor (SOR factor), where $1 < \omega < 2$. Young [1950] developed a method for determining the optimum value of ω for certain kinds of problems arising in the numerical solution of partial differential equations. Young and Gregory [1973] is a good reference.

Exercises

1. Suppose you have a relatively fast computer (a 1 µs multiply-divide time, for instance) and all the memory that you need for this problem. You need to solve a 1000×1000 system of equations that can be put into tridiagonal form. Estimate the storage required and the running time for (a) a tridiagonal-equation solver, and (b) a traditional Gaussian elimination with back-solving program. Assume for purposes of this problem that add and subtract, load and store, and housekeeping operations will require three times the multiply-divide times, so that the total running time will be four times the total times for multiply-divide operations.

2. Prove each of the following.

   a. $n(n-1) + (n-1)(n-2) + \cdots + (2)(1) = \dfrac{n^3}{3} - \dfrac{n}{3}$

   b. $1 + 2 + 3 + \cdots + (n-1) + n = \dfrac{n^2}{2} + \dfrac{n}{2}$

3. Find the inverse of

$$A = \begin{bmatrix} 1 & 1 \\ 1.01 & 1 \end{bmatrix}$$

   by performing a Gauss-Jordan reduction on A augmented by the 2×2 identity matrix. Estimate $K(A)$ using (a) the L_2 norm, (b) the L_1 norm, and (c) the L_∞ norm.

4. Repeat Problem 3 using

$$A = \begin{bmatrix} 1 & 1 \\ 1 & 1.01 \end{bmatrix}$$

Use a computer to find the solutions of the following sets of equations. (Include the condition number as part of your report.)

5. $x/1 + y/2 + z/3 = 1$
    $x/2 + y/3 + z/4 = 0$
    $x/3 + y/4 + z/5 = 0$

6. $x/1 + y/2 + z/3 + w/4 = 1$
    $x/2 + y/3 + z/4 + w/5 = 0$
    $x/3 + y/4 + z/5 + w/6 = 0$
    $x/4 + y/5 + z/6 + w/7 = 0$

7. $u/1 + v/2 + x/3 + y/4 + z/5 = 0$
    $u/2 + v/3 + x/4 + y/5 + z/6 = 0$
    $u/3 + v/4 + x/5 + y/6 + z/7 = 0$
    $u/4 + v/5 + x/6 + y/7 + z/8 = 0$
    $u/5 + v/6 + x/7 + y/8 + z/9 = 1$

8. Use a random-number generator to generate the coefficients of a fifth-order system of equations. Find at least 10 such solutions, and report on the condition numbers.

Solve the following systems of linear equations. Use TRIDIAG or PENTDIAG if appropriate.

9. $\begin{aligned} 3x - y \quad\quad\quad &= \quad 13 \\ x - 9y + z \quad\quad &= -14 \\ 2y + 7z + w &= \quad -1 \\ z - 9w &= -19 \end{aligned}$

10. $\begin{aligned} v - 2w + 3x + y + 2z &= \quad 4 \\ v + 2w - x - y \quad\quad &= \quad 0 \\ 2v + w - 3x - 2y - z &= -5 \\ v - 2w + 2x + y \quad\quad &= \quad 2 \\ v - w - 3x - 2y \quad\quad &= -6 \end{aligned}$

11. $\begin{aligned} 4u + v \quad\quad\quad\quad\quad\quad\quad &= 10 \\ u + 4v + w \quad\quad\quad\quad\quad &= \quad 0 \\ v + 4w + x \quad\quad\quad\quad &= \quad 0 \\ w + 4x + y \quad\quad\quad &= \quad 0 \\ x + 4y + z &= \quad 0 \\ y + 4z &= 10 \end{aligned}$

12. $\begin{aligned} 10u + 3v + w \quad\quad\quad\quad\quad\quad\quad &= 10 \\ 3u + 10v + 3w + x \quad\quad\quad\quad\quad &= \quad 0 \\ u + 3v + 10w + 3x + y \quad\quad\quad &= \quad 0 \\ v + 3w + 10x + 3y + z &= \quad 0 \\ w + 3x + 10y + 3z &= \quad 0 \\ x + 3y + 10z &= 20 \end{aligned}$

Problems

PROJECT PROBLEM 4.1

Write an all-purpose linear-equation solver incorporating the Gaussian elimination procedure and the back-solving procedure. The program should be interactive. It should ask the user for the order of the system to be solved and then should request the input of the A matrix by columns. This should be followed by a request for the input of the column of right-hand-side constants after a reporting of the condition number. (This allows the user to "drop out" of the program if the matrix is so poorly conditioned that the answers are questionable.)

The output should present the solution vector in an understandable manner.

PROJECT PROBLEM 4.2

1. Show that solving n problems of the form $Ax_i = b_i$, where $b_1^T = (1, 0, \ldots,$
   $0)$, $b_2^T = (0, 1, 0, \ldots, 0)$, $\ldots$, $b_n^T = (0, 0, \ldots, 0, 1)$, will give n solution
   vectors x_i, which are the columns of A^{-1} (if it exists).
2. Write a matrix inverter incorporating the Gaussian elimination and back-
   solving procedures. Provide for interactive input of n and input of A by
   columns. The output should include the condition number of A and the
   columns of A^{-1}.
3. Compute the number of multiplication-division steps required for this
   process compared to the number required for a Gauss-Jordan complete
   reduction process applied to A augmented with I.

PROJECT PROBLEM 4.3

One of the statistical procedures for fitting a polynomial to a set of experimental
data points is called the *method of least squares*. If we have n data points (x_i, y_i),
for example, we can try to fit a parabola $y = a + bx + cx^2$ to the set of points if
$n > 3$. What we do is to try to minimize

$$P(a, b, c) = \sum_{i=1}^{n} [y(x_i) - y_i]^2$$

$$= \sum_{i=1}^{n} (a + bx_i + cx_i^2 - y_i)^2$$

by proper adjustment of the values of a, b, and c. We set

$$\frac{\partial P}{\partial a} = 0 \qquad \frac{\partial P}{\partial b} = 0, \qquad \frac{\partial P}{\partial c} = 0$$

and solve the resulting system of equations:

$$na + b\sum x_i + c\sum x_i^2 = \sum y_i$$
$$a\sum x_i + b\sum x_i^2 + c\sum x_i^3 = \sum x_i y_i$$
$$a\sum x_i^2 + b\sum x_i^3 + c\sum x_i^4 = \sum x_i^2 y_i$$

Find the best-fitting parabola in the sense of least squares for the following
set of measurements:

x	y
1.000	2.420
1.500	6.041
2.000	11.610
2.500	19.127
3.000	28.625

Comment: This procedure tends to develop very poorly conditioned matrices when used with polynomials of degree higher than five or six, especially when the x_i are large, evenly spaced numbers. Project Problem 4.4 is an exercise to explore the condition numbers for such cases.

PROJECT PROBLEM 4.4

Investigate the condition numbers of the normal equations matrix for the method of least squares where the x_i are as described below. Find the condition numbers for the parabolic, cubic, quartic, and quintic curves. (The matrices for the higher-order cases, 4×4, 5×5, and 6×6, are obvious expansions of the 3×3 case developed in Project Problem 4.3.)

The x_i are calendar years 1981 through 1990, or modifications described in 2–7.

1. $x_i = 1980 + i$, $i = 1, \ldots, 10$
2. $x_i = 1980 + i - 1900 = 80 + i$, $i = 1, \ldots, 10$
3. $x_i = (80 + i)/80$, $i = 1, \ldots, 10$
4. $x_i = i$, $i = 1, \ldots, 10$
5. $x_i = 1980 + i - 1985 = i - 5$, $i = 1, \ldots, 10$
6. $x_i = i/10$, $i = 1, \ldots, 10$
7. $x_i = (i - 5)/10$, $i = 1, \ldots, 10$

What conclusions can you draw from the condition numbers you found?

PROJECT PROBLEM 4.5

A metal bar of length L, insulated on the sides, is heated to 212°F so that the temperature throughout is uniform. At $t = 0$ the ends are cooled to 32°F with ice. The initial boundary-value problem is described by

$$\frac{\partial^2 T(x, t)}{\partial x^2} = K \frac{\partial T(x, t)}{\partial t}, \qquad 0 \leq t, 0 < x < L$$

$$T(0, t) = T(L, t) = 32, \qquad 0 \leq t$$

$$T(x, 0) = 212, \qquad\qquad 0 < x < L$$

where t is the time, $T(x, t)$ is the temperature, and K is a constant called the diffusivity.

A numerical solution method involves the replacement of the partial derivatives by

$$\frac{\partial^2 T}{\partial x^2} = \frac{T(x + h, t) - 2T(x, t) + T(x - h, t)}{h^2}$$

and

$$\frac{\partial T}{\partial t} = \frac{T(x, t) - T(x, t - k)}{k}$$

where $h = L/n$, n is an integer, and k is a small interval of time. This gives a tridiagonal equation system that allows us to approximate the temperatures at a discrete set of points along the bar at discrete points in time. Generally, we replace x by the discrete values $x = ih$ and t by $t = jk$ and designate the discrete points in space-time by i, j. The temperatures are then designated by T[I, J].

Find the temperatures at 10 equally spaced points along the bar at time intervals of 2 m for about 20 m. Let L be 1 ft. The diffusivity is $K = \text{RHO} \times \text{C}/\text{COND}$, where RHO is the material density, 540 lbm/ft^3, C is the specific heat, 0.0380 Btu/lbm °F, and COND is 10.1 Btu/h ft °F.

Graphs of the temperature profiles are much more informative than tables of numbers.

Listings

LISTING OF GAUSS

```
PROCEDURE GAUSS (      N:  INTEGER;
                   VAR A:  MATRIX;
                VAR COND:  REAL;
                VAR IPVT:  INT_VECTOR);


{GAUSS triangularizes a real matrix by Gaussian elimination
and estimates the condition number of the matrix.  Use SOLVE
to complete the solution of the linear system.  It is
important that SOLVE precede GAUSS in the source code in
Turbo Pascal.}

{GAUSS was written by Craig Schons, Boulder, Colorado.}

{Input:

  N:          Order of the matrix A

  A:          Matrix to be triangularized

  A needs to be declared as a VAR of

    TYPE MATRIX = ARRAY[1..50,1..50] OF REAL;
```

B, the "right-hand-side" vector of constants, needs to be declared as a VAR of

```
TYPE VECTOR = ARRAY[1..50] OF REAL;
```

IPVT, a vector of integers, is used to keep track of row interchanges, and must be declared as a VAR of

```
TYPE INT_VECTOR = ARRAY[1..50] OF INTEGER;
```

Output:

A contains an upper triangular matrix U and a permuted version of a lower triangular matrix.

COND is an estimate of the condition of A for the linear system A*X = B. If COND = COND + 1 to machine precision, then COND is set to 1.0E+32. This assumes that A is singular.

IPVT is the pivot vector. IPVT[K] is the index of the k-th pivot row.   IPVT[N] = (-1)**(number of interchanges)

The determinant of A can be obtained on output as

```
    det(A) := IPVT[N]*A[1,1]*A[2,2]* . . . *A[N,N]   }
```


```
VAR        EK, T, ANORM, YNORM, ZNORM :    REAL;
        NMI, I, J, K, KPI, KB, KMI, M :    INTEGER;
                                    RTF :    BOOLEAN;
                                   WORK :    VECTOR;

PROCEDURE SINGULARITY;

{Exact singularity}

  BEGIN
    COND := 1.0E+32;
  END;


PROCEDURE SMALLMATRIX;

  {1 by 1 array}

  BEGIN
    COND := 1.0;
    IF A[1,1] = 0.0 THEN SINGULARITY;
  END;
```

```
PROCEDURE NORMA;

   {Compute 1-norm of A}

   BEGIN
     ANORM := 0.0;
     FOR J := 1 TO N DO
       BEGIN
         T := 0.0;
         FOR I := 1 TO N DO
           T := T + ABS(A[I,J]);
         IF T>ANORM THEN ANORM := T;
       END;
END;


PROCEDURE PIVOT;

   BEGIN
     FOR I := KPI TO N DO
       A[I,K] := -A[I,K]/T;

   FOR J := KPI TO N DO
     BEGIN
       T := A[M,J];
       A[M,J] := A[K,J];        {Swap}
       A[K,J] := T;
       IF T <> 0 THEN
         FOR I := KPI TO N DO
           A[I,J] := A[I,J] + A[I,K]*T;
         END;
END;


PROCEDURE GAUSS;

BEGIN

  FOR K := 1 TO NMI DO
    BEGIN
      KPI := K + 1;

{Find pivot}

      M := K;
      FOR I := KPI TO N DO
        IF ABS(A[I,K]) > ABS(A[M,K]) THEN M := I;
        IPVT[K] := M;
        IF M <> K THEN IPVT[N] := - IPVT[N];
        T := A[M,K];
```

```
          A[M,K] := A[K,K];
          A[K,K] := T;

{Skip step if pivot is zero}

      IF T <> 0.0 THEN PIVOT;

    END;
END;


PROCEDURE ATRANS1;

  BEGIN
    T := 0.0;
    IF K <> 1 THEN
      BEGIN
        KMI := K - 1;
        FOR I := 1 TO KMI DO
          T := T + A[I,K]*WORK[I];
      END;
    EK := 1.0;
    IF T < 0.0 THEN EK := -1.0;
    IF A[K,K] = 0.0 THEN RTF := TRUE
    ELSE
      BEGIN
        WORK[K] := -(EK + T)/A[K,K];
        K := K + 1;
      END;
END;


PROCEDURE ATRANS2;

  BEGIN
    FOR KB := 1 TO NMI DO
    BEGIN
      K := N - KB;
      T := 0.0;
      KPI := KPI;
      FOR I := KPI TO N DO
        T := T + A[I,K]*WORK[K];
      WORK[K] := T;
      M := IPVT[K];
      IF M <> K THEN
      BEGIN
        T := WORK[M];
        WORK[M] := WORK[K];
        WORK[K] := T;
      END;
    END;
```

```
YNORM  := 0.0;
  FOR I := 1 TO N DO
    YNORM  := YNORM + ABS(WORK[I]);

{Solve A*Z = Y to estimate condition}


SOLVE(N,A,WORK,IPVT);

ZNORM := 0.0;
FOR I := 1 TO N DO
  ZNORM := ZNORM + ABS(WORK[I]);

{Estimate condition}

COND := ANORM*ZNORM/YNORM;
IF COND < 1.0  THEN COND := 1.0;
END;        {ATRANS2}

BEGIN                                   {GAUSS}

  RTF := FALSE;
  IPVT[N] := 1;
  IF N = 1 THEN SMALLMATRIX
  ELSE
    BEGIN
      NMI := N - 1;
      NORMA;
      GAUSS;
```

{COND = (1-norm of A)*(estimate of 1-norm of A inverse).
Estimate is obtained by one step of inverse iteration for
the small singular vector. This involves solving two
systems of equations, (A-transpose)*Y = E and A*Z = Y where
E is a vector of +1 or -1 chosen to cause growth in Y.
Estimate = (1-norm of Z)/(1-norm of Y).}

{Solve A-transpose)*Y = E}

```
    K := 1;
    REPEAT
      ATRANS1;   {RTF = TRUE if exact singularity reached.}
    UNTIL (RTF = TRUE) OR (K > N);
    IF RTF <> TRUE THEN ATRANS2
    ELSE SINGULARITY;
  END;
END;                              {of GAUSS}
```

LISTING OF SOLVE

```
PROCEDURE  SOLVE (          N:  INTEGER;
                    VAR  A:  MATRIX;
                    VAR  B:  VECTOR;
                 VAR IPVT:  INT_VECTOR) ;
```

```
{Solution of a linear system A*X = B by "back-solving."
The matrix A must have been made triangular by the
procedure GAUSS.  Do not use SOLVE if GAUSS has detected
singularity of A.}
```

```
{Input:

  N     = Order of matrix A

  A     = Triangularized matrix received from GAUSS

  B     = Right-hand-side vector

  IPVT  = Pivot vector received from GAUSS

Output:

  B     = Solution vector X}
```

```
{Calling program must declare TYPES:

  MATRIX:    ARRAY[1..N,1..N] OF REAL

  VECTOR:    ARRAY[1..N] OF REAL

  INT_VECTOR:   ARRAY[1..N] OF INTEGER}
```

```
VAR      KB, KMI, NMI, KPI, I, K, M:      INTEGER;
                                  T:      REAL;

BEGIN           {Forward elimination}

  IF N<>1 THEN
    BEGIN
      NMI := N - 1;
      FOR K := 1 TO NMI DO
        BEGIN
          KPI := K + 1;
```

```
                M := IPVT[K];
                T := B[M];
                B[M] := B[K];
                B[K] := T;
                FOR I := KPI TO N DO
                  B[I] := B[I] + A[I,K]*T;
                END;

    {Back substitution}

    FOR KB := 1 TO NMI DO
      BEGIN
        KMI := N - KB;
        K := KMI + 1;
        B[K] := B[K]/A[K,K];
        T := -B[K];

        FOR I := 1 TO KMI DO
            B[I] := B[I] + A[I,K]*T;
      END;
    END;
    B[1]  := B[1]/A[1,1];
    END;
```

LISTING OF TRIDIAG

```
    PROCEDURE TRIDIAG(    N:   INTEGER;
                     VAR A:   VECTOR;
                     VAR B:   VECTOR;
                         C:   VECTOR;
                         D:   VECTOR);
```

{This procedure solves a tridiagonal linear system in which
 the vector D is the main diagonal of the matrix of
 coefficients, the vector A is the subdiagonal, and vector C
 is the superdiagonal. Vector B is the "right-hand-side"
 vector of constants. The vector B contains the solution
 upon return.

 The calling program must declare

     TYPE VECTOR = ARRAY[1..NUM] OF REAL;

 where NUM is the expected maximum number of elements in
 each vector (order of the tridiagonal matrix).

 The solution method is Gaussian elimination modified for
 tridiagonal systems. The computing time is on the order of
 NUM rather than NUM cubed.}

```
VAR     MULT:   REAL;
            I:    INTEGER;

BEGIN

  FOR I := 2 TO N DO
    BEGIN
      MULT := A[I-1]/D[I-1];
      D[I] := D[I] - MULT*C[I-1];
      B[I] := B[I] - MULT*B[I-1];
    END;
      B[N] := B[N]/D[N];

  FOR I := N-1 DOWNTO 1 DO
      B[I] := (B[I] - C[I]*B[I+1])/D[I];

END;
```

LISTING OF PENTDIAG

```
PROCEDURE PENTDIAG(       N:  INTEGER;
                    VAR A:  VECTOR;
                    VAR B:  VECTOR;
                    VAR C:  VECTOR;
                    VAR D:  VECTOR;
                    VAR E:  VECTOR;
                    VAR F:  VECTOR);

{PENTDIAG is a procedure to solve banded linear systems
 which have two subdiagonals and two superdiagonals in the
 coefficient matrix.  The diagonals are stored as vectors:

     E is the sub-subdiagonal
     A is the subdiagonal
     D is the diagonal
     C is the superdiagonal
     F is the super-superdiagonal
     B is the vector of right-hand-side constants

The procedure involves Gaussian elimination and back-
solving. Vector B contains the solution upon exit.

The calling program MUST declare type for the calling
program equivalents of A,B,C,D,E,F:

     TYPE  VECTOR = ARRAY[1..NUM] OF REAL;

This global declaration establishes the TYPE VECTOR in the
procedure heading.
```

Note that not all elements of A,C,E, and F are used.}

```
VAR     MULT:    REAL;
            I:    INTEGER;

BEGIN

  FOR I:= 2 TO N-1 DO
    BEGIN
      MULT  := A[I-1]/D[I-1];
      D[I]  := D[I] - MULT*C[I-1];
      C[I]  := C[I] - MULT*F[I-1];
      B[I]  := B[I] - MULT*B[I-1];

      MULT  := E[I-1]/D[I-1];
      A[I]  := A[I] - MULT*C[I-1];
      D[I+1]  := D[I+1] - MULT*F[I+1];
      B[I+1]  := B[I+1] - MULT*B[I-1];
    END;

  MULT  := A[N-1]/D[N-1];
  D[N]  := D[N] - MULT*C[N-1];
  B[N]  := (B[N] - MULT*B[N-1])/D[N];
  B[N-1]  := (B[N-1] - C[N-1]*B[N])/D[N-1];

  FOR I := N-2 DOWNTO 1 DO

    B[I]  := (B[I] - F[I]*B[I+2] - C[I]*B[I+1])/D[I];

END;
```

5

Solutions of
Nonlinear Equations

Suppose we have a certain kind of pump to transfer water from one open tank to another. The water levels in the two tanks are the same at the start, and the pump is located at the tank being emptied. To a first approximation, the pump increases the pressure in the water on its outlet side over the pressure on its inlet side by an amount given by

$$39.0 - 0.0301Q^{0.5} \quad \text{lbf/in.}^2$$

where Q is the flow rate in gallons per minute. The pressure drop in a horizontal pipe of length L (feet) due to friction losses is given by

$$\frac{0.000215 f \rho L Q^2}{D^5}$$

where f is a dimensionless constant (0.026 for water flowing in a reasonably smooth pipe), ρ is the water density (62.4 lbm/ft^3), and D is the internal diameter of the pipe in inches. If both tanks are open and at the same level, water will flow at a rate Q such that the gain in pressure in the pump is equal to the loss in pressure in the pipe. If we use a 2.50-in. pipe 200 ft long, how many gallons per minute can we pump to begin with? (As the water levels in the tanks become significantly different, the rate will decrease, of course. This is left as an exercise.)

The problem is one of solving a nonlinear equation of the form

$$ax^2 + bx^{1/2} + c = 0$$

which is not a quadratic equation. By letting $x^{1/2} = y$, we can convert the equation to one of fourth degree, for which there is a closed procedure. However, if the exponent were 0.51 instead of $\frac{1}{2}$, which could easily be the case for an empirical equation, then there would be no hope of solving except by iterative procedures. We explore some of these in this chapter. It is left as an exercise to show that the solution of the preceding problem is about 232.3 gal/min.

The solution of nonlinear equations is a difficult problem. Part of the difficulty is that each problem is different and there is no generally recognized "best" method of solution. Furthermore, there is not an extensive general theory that can be applied, as there is, for example, in the case of systems of linear algebraic equations. Only in the case of a single polynomial equation is there much theory that is very helpful.

However, if something is known about the nature of the problem, there are powerful methods of finding roots provided we know the approximate locations of those roots. We examine several of these methods in this chapter. First it is necessary to look at iterative processes and the nature of their convergence.

5.1 Iteration

Iteration stems from a Latin word that means to say again, or to repeat. Iterative processes, if they converge, produce a set of values, say $\{x_n\}$, that are successively closer to the desired solution. By convergence we mean the following:

Given an $\epsilon > 0$, arbitrarily small, there exists an $N(\epsilon)$ such that $|x - x_n| < \epsilon$ whenever $n > N$, where x is the point of convergence.

We often use the abbreviated notation $x_n \rightarrow x$ to stand for the notions in the definition. Table 5.1 shows the successive iterates x_n obtained during the process of finding the square root of 81 by an iterative process.

Table 5.1

Iteration	Value of X
1	20.
2	12.025
3	9.380483368
4	9.007716425
5	9.000003305
6	9.000000000

Unfortunately, not all iterative processes converge as rapidly as this. One of the insights we seek in this chapter is why some processes converge rapidly and others do not or do not converge at all.

5.2 Interval Halving

Interval halving, often called the *method of bisection*, is the simplest of iterative methods. It is a kind of "divide-and-conquer" algorithm similar to the well-known method for catching a lion in the desert:

> To catch a lion in the desert, simply erect a north-south fence across the desert and then determine in which half the lion is to be found. Subdivide that half with another fence and make another determination of location. Continue this process until the lion is as localized as desired. (In case you have a two-dimensional desert it is desirable to alternate between north-south and east-west fences.)

To begin, we must first bracket the root between two values, x_a and x_b, where $x_a < x_b$. If the equation we are trying to solve is of the form $f(x) = 0$ and if $f(x)$ is a continuous function, we know that there will be at least one root in the interval if the signs of $f(x_a)$ and $f(x_b)$ are different. There is no foolproof method for finding such initial pairs of points bounding the roots. One practical method that will work in many cases is simply to step along the x axis by unit steps, testing the sign of $f(x)$ as we go. However, we will often miss pairs of roots that are less than one unit apart; the closer together they are, the less likely they will be to straddle a unit point (as compared to falling between unit points). However, nothing helps more than a careful sketch in locating starting values. In Figure 5.1 there is a root at x_1 that satisfies the sign criterion, but at x_2 there is a double root for which the sign criterion is not satisfied. Figure 5.2 shows a case where three roots are located between x_a and x_b. It is also essential to the method that the function be continuous. Figure 5.3 shows the graph of a function with sign changes at x_1 and at x_2 but with no roots at those points.

Assume that we have located two points x_a and x_b such that the sign of $f(x)$ differs at the two points. We now repeat (iterate) the following:

1. Let $x_c = (x_a + x_b)/2$.
2. If $f(x_c) = 0$, stop; x_c is the root.

FIGURE 5.1

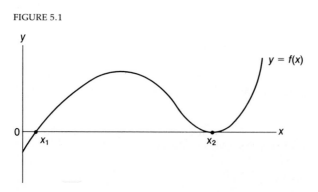

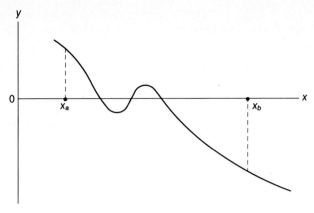

FIGURE 5.2

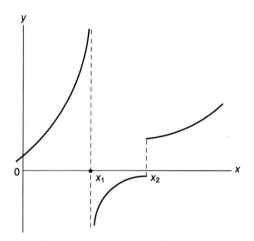

FIGURE 5.3

3. If the signs of $f(x_c)$ and $f(x_a)$ agree, then replace x_a by x_c; otherwise replace x_b by x_c.
4. Return to Step 1.

This algorithm can be improved by adding a test to terminate the process. Replace Step 4 by:

4. If $\mathrm{ABS}(x_a - x_b) < \mathrm{EPS}$, then stop with $x_c = (x_a + x_b)/2$ as the root; otherwise return to Step 1.

The choice of a value for EPS is dictated by considerations that we will examine later in the chapter. Suffice it to say now that this is one of several methods of stopping the iteration.

This method is slow but reliable. In contrast to some of the faster and more sophisticated methods, there is no place at which it can go terribly wrong as long as $f(x)$ is continuous and as long as you can find x_a and x_b such that $f(x_a) \cdot f(x_b) < 0$. The rate of convergence is said to be linear. That is, the uncertainty in the root at one stage is a fixed fraction (one-half) of the uncertainty at the previous stage. It is easy to see how many steps will be required to obtain a root with desired accuracy.

Suppose we start with x_a and x_b one unit apart and suppose we wish to find an estimate of the root that is in error by no more than 10^{-6}, for instance. We thus need to solve for n in the following inequality:

$$(x_b - x_a)(\tfrac{1}{2})^n = (\tfrac{1}{2})^n < 10^{-6}$$

which yields $n > 6/\log 2 = 19.93$. Thus 20 steps will suffice.

The stopping criteria can offer a problem. The usual criterion is to stop when the interval of uncertainty containing the root is a small multiple of the machine epsilon times the absolute value of the estimated root. The simple test $|x_a - x_b| < \text{EPS}$ requires that we choose EPS wisely; unfortunately, the value of EPS depends upon the magnitude of the root. A better test is to stop when $|x_a - x_b| < \text{EPS} * |x_a|$ where EPS is a small multiple of the machine epsilon and is independent of the magnitude of the root. However, if $f(x)$ has a nearly vertical graph at the root, then $f(x)$ may not be close to zero. It may then pay to test both endpoints in f to see which gives the smaller functional value. See Figure 5.4. Simply accepting a small value for $f(x)$ is not a satisfactory stopping criterion because $f(x)$ can have a horizontal tangent at a root. This means that a large range of values of x may satisfy $|f(x)| < \epsilon$. See Figure 5.5. The safest procedure is to require that both stopping criteria be satisfied before iteration is terminated.

FIGURE 5.4

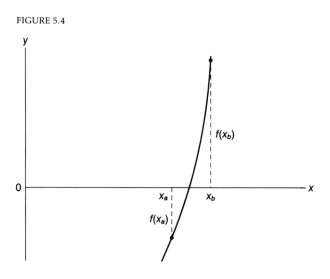

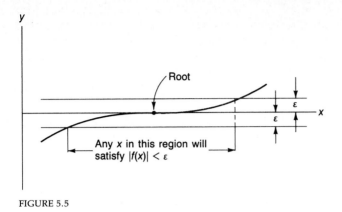

FIGURE 5.5

5.3   Method of Successive Substitution

Sometimes called the *method of iteration,* successive substitution is often a useful method, although it is usually slow. We wish to look at it because it is a simple method and has an easy convergence proof.

We start by arranging the equation to be solved in the form $x = F(x)$ and make an initial guess x_0 at the value of the root. We then go through the steps

$$x_1 = F(x_0)$$
$$x_2 = F(x_1)$$

$$\cdot$$
$$\cdot$$
$$\cdot$$

$$x_{n+1} = F(x_n)$$

If the process converges, the successive values of x approach the desired root, which may be expressed in the form

$$\lim x_n = x$$

where x is the root of $x = F(x)$.

Example 5.1

Suppose we wish to find the two roots $x = 2$ and $x = 3$ of the quadratic equation

$$x^2 - 5x + 6 = 0$$

(*Remark:* This is certainly not the way to solve a quadratic equation, but it is instructive to see how the method can be applied to a simple example.)

First we solve for one of the x's:

$$x = \frac{x^2 + 6}{5}$$

If we choose $x_0 = 1.0$, we then find

$$x_1 = \frac{1^2 + 6}{5} = 1.4$$

$$x_2 = 1.592$$

$$x_3 = 1.7068928$$

.

.

.

$$x_{25} = 1.998430610$$

It appears that x_n is approaching $x = 2$, albeit rather slowly.
Suppose we now try $x_0 = 4$. We then find the following sequence:

$$x_1 = 4.4$$

$$x_2 = 5.072$$

$$x_3 = 6.3450368$$

.

.

.

$$x_8 = 174{,}753.3656 \qquad \square$$

Clearly, the sequence $\{x_n\}$ is divergent. How can we tell whether convergence will occur? The answer to that question is not always easy, and it involves understanding the condition for convergence.

First we assume that $F(x)$ is continuous and satisfies a condition that for all x, y in some region on the real axis,

$$|F(x) - F(y)| < m|x - y|$$

where $0 < m < \infty$. Now assume a starting value x_0 lying within the region. Then we have

$$|F(x) - F(x_0)| = |x - x_1|$$

where x is the (unknown) root. But

$$|F(x) - F(x_0)| < m|x - x_0|$$

because of the conditions we imposed. Hence we have

$$|x - x_1| = |F(x) - F(x_0)| < m|x - x_0|$$

Likewise,

$$|x - x_2| = |F(x) - F(x_1)| < m|x - x_1| < m^2|x - x_0|$$

.

.

.

$$|x - x_n| = |F(x) - F(x_{n-1})| < m|x - x_{n-1}| < m^n|x - x_0|$$

Thus we see that if $0 < m < 1$, then $m^n \to 0$ as $n \to \infty$, and $|x - x_n|$ can be made as small as we wish by making n sufficiently large.

The condition

$$|F(x) - F(y)| < m|x - y|$$

is equivalent to

$$\frac{|F(x) - F(y)|}{|x - y|} < m$$

Note that the limit

$$\lim_{y \to x} \frac{F(x) - F(y)}{x - y}$$

defines the derivative $F'(x)$. Any function $F(x)$ that is continuous in the neighborhood of the root of $x = F(x)$ and that satisfies the condition $|F'(x)| < 1$ at the root x has a neighborhood of x for which $|F'| < 1$ everywhere. If we choose a starting value in this neighborhood and no iterates go outside of the neighborhood, then there will be an iterative solution that will find x.

It is clear from the derivation of the condition for convergence that the size of m has an effect on the rate of convergence. Near the root the error at each step of the iteration is about m times as great as the error on the previous step. Thus if m is very small, convergence will be much faster than if m is near 1.

Example 5.2

If $F(x) = (x^2 + 6)/5$, $F'(x) = 2x/5$. The region for which the derivative condition holds is

$$\left|\frac{2x}{5}\right| < 1$$

or

$$\frac{-5}{2} < x < \frac{5}{2}$$

At the root $|F'(x)| = \frac{4}{5}$, so we can expect convergence to be slow. Also note that $x_0 = 4$, the starting value for the iteration for the other root, and the root $x = 3$ are outside the convergence region.

If $F(x) = (5x - 6)^{1/2}$, which is what we get if we solve for the other x in the quadratic equation $x^2 - 5x + 6 = 0$, then we find that $|F'(x)| = \frac{5}{6}$ in the neighborhood of the root $x = 3$.

Figure 5.6(a) shows the behavior of the iteration process for this example. Figure 5.6(b) shows the behavior when we try to solve for the root $x = 3$, using $F(x) = (x^2 + 6)/5$. You are encouraged to try other starting values near the root $x = 3$ for both cases. □

5.4   Rate of Convergence

In the previous section we saw that if $|F'(x)| < 1$ in the neighborhood of the root and if the starting value was included in this neighborhood, the iteration process converged. For the root $x = 2$, we saw that

$$|2 - x_n| \cong \left(\tfrac{4}{5}\right)|2 - x_{n-1}|$$

If we call $|2 - x_n|$ the error at the nth step, e_n, we then have the relation

$$e_n \cong me_{n-1}$$

where $m = \frac{4}{5}$ in this example. It will take many steps to reduce the error to an acceptable value. Suppose x_0 is about one unit away from x, so that $|x - x_0| \cong 1$, and suppose we wish to have no more error than 10^{-6} at the termination. Then

$$|x - x_n| \leqslant m^n |x - x_0|$$

which for $e_n = 10^{-6}$ means $n = 6/\log\left(\tfrac{5}{4}\right) = 62$ iterations. Processes in which $e_n \cong me_{n-1}$ are said to be linearly convergent. Other kinds of convergence are possible, such as quadratic convergence, in which $e_n \cong m(e_{n-1})^2$. In this type of convergence, not nearly so many steps are required, in general. An example is Heron's method for finding the positive square root of a positive number, which was illustrated early in the chapter (Table 5.1). That process required only a few steps to obtain a highly accurate answer, even though the starting value was not very close to the root. Unfortunately, the more sophisticated methods are more susceptible to disaster, as we shall see.

5.5   Terminating the Iterative Process

In human terms we think of iterating until the answer is "good enough." That is not a satisfactory criterion to present to the computer, so we need to develop precise conditions which must be met by the process.

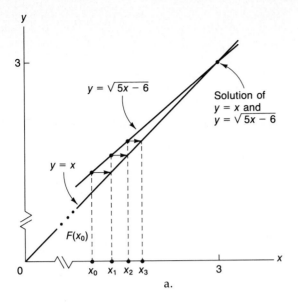

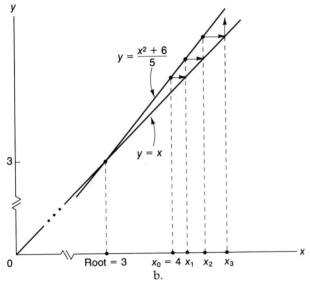

FIGURE 5.6

One common way of terminating an iterative process is to stop when $|x_n - x_{n-1}|$ is sufficiently small. Although this is satisfactory in some circumstances, it is not so in others. For example, suppose the value of $|x|$, where x is the root, is on the order of 1. Then for a 10-decimal machine, the value of $|x_n - x_{n-1}|$ when we are "happy" with the answer is likely to be near the machine epsilon, say on the order of 10^{-10}. This will be entirely unsatisfactory if $|x|$ is on the order of 10^{20}. In that case we will have to terminate the iteration when $|x_n -$

x_{n-1}| is on the order of 10^{10}. Attempting to use a smaller stopping epsilon might lead to an endless loop.

A safer stopping criterion is to require that

$$|x_n - x_{n-1}| < \text{EPS} \cdot |x_n|$$

where EPS is two to four times the machine epsilon. Note that this is a relative rather than an absolute criterion, which works in floating point arithmetic regardless of the size of the root.

5.6   Newton's Method

The idea behind Newton's method is to replace the function $f(x)$ in the equation to be solved ($f(x) = 0$) by a simpler function $p(x)$ that approximates the function $f(x)$ in the vicinity of the root in order to solve the simpler equation $p(x) = 0$. The simplest useful function is a linear function, which for Newton's method is chosen to be the tangent line to $f(x)$ at a point whose x coordinate is an initial approximation to the root. The point where the tangent line crosses the x axis is usually a better approximation to the root. Figure 5.7 shows a sequence of such tangent-line approximations of $f(x)$ and how they produce a convergent set of estimates of the root of $f(x) = 0$.

In Figure 5.7 we see that the right triangle with vertices $(x_1, 0)$, $(x_2, 0)$, and $(x_1, f(x_1))$ has a hypotenuse with slope $f'(x_1)$. The slope is also $f(x_1)/(x_1 - x_2)$. If we equate these two expressions and solve for x_2, we find that

$$x_2 = x_1 - \frac{f(x_1)}{f'(x_1)}$$

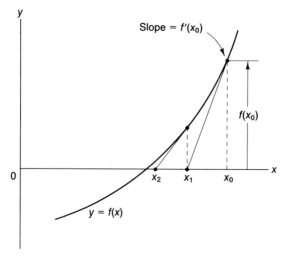

FIGURE 5.7

or, in general,

$$x_{n+1} = x_n - \frac{f(x_n)}{f'(x_n)}$$

Example 5.3

Solve $N - x^2 = 0$ to find the positive square root of N, $N > 0$.

$$f(x) = N - x^2, \qquad f'(x) = -2x$$

Thus

$$x_{n+1} = x_n - \frac{(N - x_n^2)}{(-2x_n)}$$

$$= \frac{x_n + N/x_n}{2}$$

This is Heron's method for finding square roots. It was used to produce Table 5.1 near the beginning of the chapter. □

To show that Newton's method is quadratically convergent, we first write a Taylor formula expression for $f(a)$, where a is the root, about a point of expansion x_n:

$$f(a) = 0 = f(x_n) + f'(x_n)(a - x_n) + \frac{f''(z)(a - x_n)^2}{2}$$

where

$$a < z < x_n \quad \text{or} \quad x_n < z < a$$

Since $e_n = x_n - a$, we can write

$$0 = f(x_n) - e_n f'(x_n) + \frac{f''(z)e_n^2}{2}$$

If we solve for $f(x_n)/f'(x_n)$ from this equation and substitute into the Newton formula, modified to read

$$x_{n+1} - a = x_n - a - \frac{f(x_n)}{f'(x_n)}$$

we find that

$$e_{n+1} = \frac{e_n^2 \times \frac{1}{2}f''(z)}{f'(x_n)}$$

If $f'(x)$ and $f''(x)$ are well behaved in the neighborhood of the root (f'' does not increase or decrease without limit and f' does not approach zero as x approaches the root), then we have quadratic convergence.

Note that if f' is zero at the root the rate of convergence is not quadratic. This happens at multiple roots.

Example 5.4

Let $f(x) = (x - 1)^2 = 0$, which has a double root at $x = 1$. Convergence is very slow:

$$x_0 = 0.5$$
$$x_1 = 0.75$$
$$x_2 = 0.875$$

.

.

.

$$x_{10} = 0.9995117188$$

.

.

.

$$x_{20} = 0.9999995232$$

This is obviously not quadratic convergence!   □

What can be done about multiple roots? There are preventative measures available that permit rapid convergence in spite of the multiplicity of roots. One such method is to use three terms of the Taylor expansion of the function $f(x)$ to derive a new formula to replace Newton's formula.

A preferable method [Ralston, 1978] is to recognize that if a function has a zero of multiplicity m (that is, if $f(x) = 0$ has m equal roots r), then $f(x)$ has a Taylor expansion of the form

$$f(x) = a_m(x - r)^m + a_{m+1}(x - r)^{m+1} + \cdots$$

Note that $f'(x) = 0$ has a root with multiplicity $m - 1$. Hence $f(x)/f'(x) = 0$ has a single root $x = r$, and it should be possible to solve $f(x)/f'(x) = 0$ by Newton's method to obtain an answer with quadratic convergence. The following example shows some of the problems.

Example 5.5

Solve $e^{x^2} = 1$ for x. The solution is in the field of reals with $x = 0$ as a double root (the solution is really that of $x^2 = 0$). If we start Newton's method with $x_0 = 1$, we obtain

$x_1 = 0.6839397206$

$x_2 = 0.4108128437$

$x_3 = 0.2218040674$

$x_4 = 0.1135858677$

$x_5 = 0.0571577275$

This is obviously linear convergence caused by the double root at $x = 0$.   □

Now suppose we consider $f(x)/f'(x) = 0$. Each zero of $f(x)$ is also a zero of f/f'. However, each multiple root of $f = 0$ will occur only once for $f/f' = 0$. Consequently the Newton method for $u(x) = f(x)/f'(x)$ ought to converge quadratically for any case of multiple roots (but there is no guarantee that there may not be problems).

Example 5.6

Let us return to the previous example, where we tried to solve $e^{x^2} = 1$. If we let $f(x) = e^{x^2} - 1 = 0$ and $u(x) = f(x)/f'(x)$, then $u(x)/u'(x) = x(e^{x^2} - 1)/(2x^2 + 1 - e^{x^2})$. There is no guarantee that the region of convergence is the same; in fact, the $f(x)$ sequence converges from $x = 1$, but the $u(x)$ sequence diverges from $x = 1$. With a starting point of $x_0 = 0.5$, the sequence $u(x)$ gives

$x_1 = -0.1575436154$

$x_2 = 0.0039927459$

$x_3 = -0.0000000629$

$x_4 = $ (blows up!)   □

The cause of the disaster was division by zero; the denominator of $u(x)/u'(x)$ gave a machine zero for x_3. This is an example of the price paid for speed and sophistication in many problems.

A way of circumventing this disaster is to factor out the original $f(x)$ on the right-hand side of the iteration equation so that it reads

$$x_{n+1} = x_n - f(x_n)\left[\frac{1}{f'(x_n) \cdot u'(x_n)}\right]$$

The program is then modified to do two things:

1. Preserve the old value of the quantity in square brackets each time before it is recomputed.
2. Test all denominators for near-zero conditions. If a division overflow is about to occur, the old value of the quantity in square brackets is used instead of trying to compute a new value.

Newton's method will converge even if the slope value in the iteration formula is not highly accurate. As we shall see in the next section, such processes converge less rapidly than quadratically. With the suggested modification of the program (use of the old value of the quantity in the square brackets when the denominator fell below 10^{-20} in magnitude), the process terminated with $x_4 = 0.0$, which meant that $|x_4|$ converted to the machine zero.

5.7   Two-point Formulas

In some cases evaluating $f(x)$ and $f'(x)$ at each step may be very costly. In such cases it is useful to have two-point formulas, in which we approximate the slope of the tangent line in Newton's method by the slope of a secant line determined by two points on the graph of $f(x)$. Each step of the iteration then requires only one function evaluation and may thus save as much as half the computer time.

In the secant method we let $f'(x_n)$ be approximated by

$$\frac{f(x_n) - f(x_{n-1})}{x_n - x_{n-1}}$$

Then

$$x_{n+1} = x_n - \frac{(x_n - x_{n-1})f(x_n)}{f(x_n) - f(x_{n-1})}$$

The algorithm for using this formula requires that we have two initial approximations to the root, x_0 and x_1. We also need to calculate $f(x_0)$ and $f(x_1)$. Let $D_{new} = f(x_1) - f(x_0)$, and let $n = 1$.

1. Compute $x_{n+1} = x_n - (x_n - x_{n-1})f(x_n)/D_{new}$.
2. If $|x_{n+1} - x_n| < e_1 \cdot |x_{n+1}|$ and $|f(x_{n+1})| < e_2$, then terminate with x_{n+1} as the approximate root. Otherwise proceed to Step 3.
3. Save D_{new} as D_{old}. Calculate the new value of D as $D_{new} = f(x_{n+1}) - f(x_n)$. If $|D_{new}| < 10^{-20}$, then replace D_{new} by D_{old}. Replace x_{n-1} by x_n, x_n by x_{n+1}, and $f(x_n)$ by $f(x_{n+1})$. Let $n = n + 1$ and return to Step 1.

You are encouraged to work the exercise in the problems in which the secant-method formula is derived from a geometrical construction.

There is a temptation to rewrite the secant-method equation to combine x_n with the larger expression to obtain a nice, symmetrical formula. Unfortunately, this can lead to poor accuracy in computation because of cancellation errors. It is usually better to compute something as a small change to a good approximation than to compute it from the beginning.

It can be shown [Dahlquist and Bjorck, 1974] that the order of convergence of the secant method is about 1.62; that is,

$$e_{n+1} = Ce_n^{1.62}$$

While this is not as rapid as Newton's method, it may be more efficient computationally because it requires only one function evaluation per step. This is particularly true if $f(x)$ and $f'(x)$ are complicated functions.

Example 5.7

Let us repeat the example of finding the square root of 81 using the secant method. For the two required starting values, we will use 20 and 12.025, so that we can compare the speed of convergence of this method with that of Newton's method.

The secant equation reduces to

$$x_{n+1} = x_n + \frac{N - x_n^2}{x_n + x_{n-1}}$$

The sequence of iterates is:

$x_0 = 20.0$

$x_1 = 12.025$

$x_2 = 10.03903201$

$x_3 = 9.142452287$

$x_4 = 9.007716425$

$x_5 = 9.000060563$

$x_6 = 9.000000026$

$x_7 = 9.000000000$ □

5.8 Accelerating Convergence

There are generalized procedures, which depend upon properties of the convergence behavior, that can be used to speed convergence. For example, if the differences between successive iterates tend to decrease linearly, then a better estimate of x_{n+1}, called x'_{n+1}, can be obtained from three successive iterates:

$$x'_{n+1} = x_{n+1} - \frac{(x_{n+1} - x_n)^2}{x_{n+1} - 2x_n + x_{n-1}}$$

This procedure, known as Aitken's method, can be used to produce a sequence of $\{x'_n\}$ from a sequence $\{x_n\}$. Likewise, this new sequence can be used to produce another sequence $\{x''_n\}$, which converges still faster. Each succeeding sequence will be shorter than the previous one by two, and the process terminates when there are only one or two terms. This (these) value(s) can be tested in the original equation. Even if the result is not yet satisfactory, there is often a remarkable gain in accuracy.

Example 5.8

Table 5.2 shows the values of x_n in the solution of

$$(x - 1)^3 = 0$$

by Newton's method. Because of the triple root, the process converges only linearly with $e_{n+1} \cong \left(\frac{2}{3}\right)e_n$. The Aitken extrapolation values are shown in the right-hand column. Except for round-off errors, these values would all be 1.0.

Table 5.2 Aitken's Extrapolation for Slowly Converging Sequence

Slowly Converging Sequence	Aitken's Extrapolation
0.5	
0.6666666667	
0.7777777778	0.9999999998
0.8518518519	1.000000000
0.9012345679	0.9999999995
0.9341563786	1.000000000
0.9561042524	1.000000000
0.9707361683	1.000000000
0.9804907788	0.9999999992

5.9 Practical Considerations

What kinds of things can go wrong using a more powerful method such as Newton's method? For one thing, Newton's method can be used to find only real roots unless a complex starting value is used. For another, it may find the wrong root if the starting value is not chosen with care, as shown in Figure 5.8. Sometimes it can "get stuck," or "hang," for a long time in situations such as that shown in Figure 5.9. Similar pitfalls can be encountered with the secant method, although the details of the hanging process are different. It is probably desirable to equip any computer program using an iterative root-finding method with a

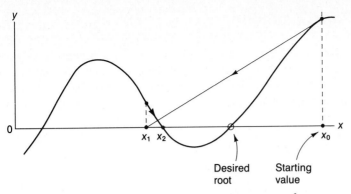

FIGURE 5.8   Poor starting value causes convergence to the
wrong root.

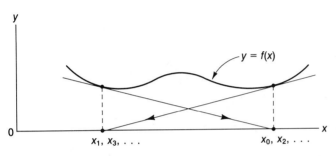

FIGURE 5.9

limit on the number of iterations that will be allowed before a new starting value
is tried.

For large and complicated problems, it may be very difficult, if not impossible, to obtain the derivative $f'(x)$ for use in Newton's method. An example of
this is a problem in which x is a parameter in a definite integral that must be
evaluated numerically. This will mandate the use of the secant method (with a
numerical integration necessary for every iteration). In very large problems it is
sometimes worthwhile to avoid the evaluation of $f'(x)$ at every step. Near the
root, high accuracy of $f'(x)$ is usually not critical. Although the Newton process
reduces to linear iteration when $f'(x)$ is fixed, the iteration constant C in

$$e_{n+1} = Ce_n$$

is very small if the root is not multiple.

You are encouraged to work at least one problem using Newton's method
and the secant method on a calculator in order to get a feel for the nature of the
convergence process.

5.10   Polynomial Equations

Polynomial equations represent an important subset of the general class of non-linear equations. The methods that we have looked at previously may be used for finding real roots and, with suitable modifications, for finding all roots, real or complex. However, because of special properties, polynomial equations have some methods of their own.

First, let use review a few properties of and facts about polynomials that may be useful. The first concerns the number of roots and the degree of the equation. A theorem from elementary algebra assures us that the number of roots equals the degree of the polynomial equation. (We must be sure to count multiple roots the number of times they occur.) Furthermore, those roots may be real or complex. If the equation has real coefficients, then the complex roots must occur in complex conjugate pairs.

For any polynomial of the nth degree,

$$p_n(x) = a_0 x^n + a_1 x^{n-1} + \cdots + a_{n-1} x + a_n$$

and for $z = x + iy$, a general complex variable, it is clear that

$$p_n(x) = u_n(x, y) + iv_n(x, y)$$

where u_n and v_n are nth-degree polynomials in x and y with real coefficients. If $p_n(x) = 0$, then it follows that $u_n(x, y) = 0$ and $v_n(x, y) = 0$. Thus if $x + iy$ is a root, then $x - iy$ must also be a root. This is because

$$p_n(x - iy) = u_n(x, y) - iv_n(x, y)$$

and both $u_n(x, y)$ and $v_n(x, y)$ are zero.

Descartes' rule of signs is often helpful in determining the number of each kind of roots to be expected. The rule states that the number of positive, real roots for a polynomial equation (with real coefficients) is equal to the number of changes of sign of the coefficients (when the equation is ordered on the powers of the variable) or an even number fewer.

Example 5.9

$$f(x) = x^4 - 2x^3 - 3x^2 - 5 = 0$$

This equation will have exactly one positive root because the coefficients show only one sign change (in moving from the x^4 to the x^3 term). However,

$$g(x) = x^4 - 2x^3 + 3x^2 - 5 = 0$$

has three changes of sign and so may have three positive real roots or one positive real root and two complex roots.   □

from which it can be found that

$$R = b_{n-1} \qquad S = b_n + r b_{n-1}$$

$$\frac{\partial R}{\partial r} = -c_{n-2} \qquad \frac{\partial S}{\partial r} = -c_{n-1} - r c_{n-2} + b_{n-1}$$

$$\frac{\partial R}{\partial s} = -c_{n-3} \qquad \frac{\partial S}{\partial s} = -c_{n-2} - r c_{n-3}$$

Thus if we define $b_{-1} = c_{-1} = b_{-2} = c_{-2} = 0$, we can generate the b's from the set of a's and the c's from the set of b's, given initial values of r and s. We then solve the linear system

$$c_{n-2}(r' - r) + c_{n-3}(s' - s) = b_{n-1}$$

$$(c_{n-1} - b_{n-1})(r' - r) + c_{n-2}(s' - s) = b_n$$

for a new estimate (r', s') of (r, s). This process is repeated until (we hope) convergence occurs. If and when it does occur, the polynomial factor

$$x^2 + rx + s$$

can be set to zero and the resulting equation can be solved by formula for a pair of roots. The other polynomial,

$$b_0 x^{n-2} + b_1 x^{n-3} + \cdots + b_{n-2}$$

is then the deflated polynomial, which can be set to zero and solved for the remaining roots.

In the procedure BAIRSTOW, the b and c coefficients are generated from the a coefficients and the 2×2 set of linear equations is solved for improved values of r and s. When convergence occurs, the values of b_{n-1} and b_n should be very close to zero (ideally, exactly zero). When the roots have been calculated and the deflated polynomial is ready to be treated, the b's from 0 to $n - 2$ are transferred to the array occupied by the a's, and the value of n is reduced by 2. If the deflated polynomial is only of first or second degree, the linear- or quadratic-equation solver is called and the program is stopped.

If convergence fails to occur within 30 iterations, the starting values of r and s are changed, and the iteration process is started again. Twelve such tries are permitted before the procedure reports failure and halts.

It is left as a project problem to modify the procedure to remove the print statements.

5.13   IMSL Routine ZBRENT

The FORTRAN routine ZBRENT finds a root of $f(x) = 0$, where $f(x)$ is a continuous function that changes sign on the interval in which the root is sought. It uses Brent's algorithm [Brent, 1971], which involves a combination of interval bisection and linear or inverse quadratic interpolation. It is at least as fast as a linearly convergent process.

The call to ZBRENT is

CALL ZBRENT (F, EPS, NSIG, A, B, ITMAX, IER)

where F is the name of the externally declared, user-supplied, function $f(x)$. EPS is the first convergence criterion number. If $|f(x)| <$ EPS, then x will be accepted as the root. NSIG is the integer value associated with the second convergence criterion, discussed shortly. A and B are the bounds on the interval in which the root will be sought (F (A) $\cdot$ F (B) $<$ 0 is required). The root, when found, will be returned as B. ITMAX is a user-supplied (integer) upper limit on the number of iterations permitted. IER is an integer error flag with a normal termination value of zero. If ITMAX is reached, IER will be set to 129. If F (A) and F (B) agree in sign, IER will be set to 130. On a normal exit from ZBRENT, the value of ITMAX is the number of iterations required to find the solution.

The second convergence test requires that

$$|x_{j-1} - x_j| < |x_j| \cdot 10^{-\text{NSIG}}$$

which means that x_j is chosen as the root when it agrees with x_{j-1} to NSIG or more decimal digits. Note that if either convergence criterion is satisfied, then x_j will be transferred into B and the return will be made with IER = 0.

5.14   IMSL Routine ZRPOLY

The FORTRAN routine ZRPOLY finds (or attempts to find) the roots of a polynomial equation with real coefficients. It requires the use of complex arithmetic.

The call to ZRPOLY is

CALL ZRPOLY (A, N, Z, IER)

where A is the real array of the polynomial coefficients, N is the (integer) degree of the polynomial, Z is the complex array of roots, and IER is an integer error flag. The polynomial is of the form

$$A_1 Z^N + A_2 Z^{N-1} + \cdots + A_N Z + A_{N+1}$$

Array A must have dimension of N + 1 (or greater) with the (nonzero) leading coefficient in A (1). The array Z must be complex and must have dimension of at least N.

On a normal return from ZRPOLY the flag IER should be zero. If not, the contents of IER indicate the following:

IER = 129:   N > 100 or N < 1
IER = 130: $A_1 = 0$

IER = 131:   Fewer than N roots were found. In
                    that case, the missing roots are
                    represented in Z by machine
                    infinity.

The method used in ZRPOLY is the Jenkins and Traub three-stage algorithm
using quadratic iteration [Jenkins and Traub, 1970].

Exercises

1. Apply Aitken's method to the first few terms of the ordinary Newton's method solution for the double root of

   $$f(x) = (x - 1)^2 = 0.$$

2. Show that the iteration

   $$x_{n+1} = 2x_n - vx_n^2$$

   can be used to compute $x = 1/v$ without the use of division. (Show that the iteration
   is convergent when x_n is close to $1/v$, and show that $x = 1/v$ satisfies the iteration
   equation.) This method was (and may still be) used on machines with no division
   hardware in preference to programmed long division.

3. With the aid of an appropriate sketch, derive the equation for the secant method. Use
   geometrical constructions and ratios of sides of similar triangles.

4. Make an appropriate sketch and use geometrical methods to derive the equation for
   Newton's method. Do this for all four combinations of $f'(x) > 0$ at the root, $f'(x) < 0$
   at the root, $x_n > x$, and $x_n < x$, where x is the root. Show that all four derivations
   give the same formula.

5. Sketch $y = x^2$ and $y = x$ on the same set of axes. Show graphically why the equation
   $x^2 - x = 0$ cannot be solved by the method of successive substitution using $x_{n+1} = x_n^2$ for the root $x = 1$ but that it will converge to the root $x = 0$ with an appropriate
   starting value. How can the problem be modified to find the root $x = 1$ by successive
   substitution?

6. Use the next-to-last entries in Table 5.1 to estimate the value of the constant of pro-
   portionality C in $e_{n+1} = Ce_n^2$. How far up the table can you go before the value changes
   significantly?

7. Show that if n is odd,

   $$x^n - 1 = (x - 1)(x^{n-1} + x^{n-2} + x^{n-3} + \cdots + x + 1)$$

   Now solving $x^n - 1 = 0$ reduces to finding the n nth roots of unity, which are

   $$\cos\left(\frac{2k\pi}{n}\right) + i \sin\left(\frac{2k\pi}{n}\right), \qquad k = 0, 1, \ldots, n - 1$$

   a. Use this to check the machine solutions of

   $$x^6 + x^5 + x^4 + x^3 + x^2 + x + 1 = 0$$

b. Create a similar result and use it to check the machine solutions of

$$x^8 - x^7 + x^6 - x^5 + x^4 - x^3 + x^2 - x + 1 = 0$$

8. Derive the Newton's method equation with the aid of a Taylor expansion truncated to two terms. Replace $f(x)$ in $f(x) = 0$ by

$$f(x) = 0 = f(x_n) + f'(x_n)(x - x_n) + \cdots$$

and solve for x, which is a presumably better approximation x_{n+1}.

9. Repeat Exercise 8, but include the $f''(x_n)$ term. Try this method on a simple equation with an imaginary root. Also try the ordinary Newton's method, but give it a complex starting value close to the known root.

10. Derive the equation for Aitken's method using the condition stated in the example. (*Hint:* It will be necessary to add and subtract certain terms to create the form shown in that section.)

11. Find the value of a such that a horizontal line $y = a$ through the graph of $y = \sin x$ divides the area enclosed by $y = \sin x$ and $y = 0$ between $x = 0$ and $x = \pi$ into two equal parts.

12. Use the secant method to derive an equation equivalent to the one for Heron's method. Calculate the square root of 81 with this formula. Use the values of x_0 and x_1 from iterations 2 and 3 in Table 5.1 as starting values.

13. A certain nonlinear spring develops a restoring force given by $F = k_1 x + k_3 x^3 + k_5 x^5$, where F is in pounds force and x is in feet. The values of the constants are

$$k_1 = 10.5 \text{ lbf/ft}$$
$$k_3 = 1.20 \text{ lbf/ft}^3$$
$$k_5 = 0.0235 \text{ lbf/ft}^5$$

A 25.0-lb weight is placed on the spring. How far will it compress?

14. The specific gravity of a hollow, spherical, metal ball is two-thirds that of water. In terms of its radius, how deep will the ball float in the water?

15. Solve $x - 2e^{-x} = 0$ on a programmable calculator using (a) the method of successive substitution and (b) Newton's method.

16. Solve $x + \ln x - \ln 2 = 0$ on a programmable calculator using (a) the method of successive substitution and (b) Newton's method.

17. Find all the real roots of each equation.
   a. $x^3 + 8x^2 - 15x - 54 = 0$
   b. $x^4 - 2x^3 - 26x^2 - 18x - 315 = 0$
   c. $x^3 + 6x - 27 = 0$

18. Find all the real roots unless otherwise specified:
   a. $2x - 20 \sin x + 1 = 0$
   b. $x + e^{-2x} = 2$
   c. $x = \tan x$ (all roots for which $0 \leq x \leq 15$)
   d. $x = e^{-3x}$
   e. $x^2 - 3x - 1/(x + 1) = 0$

19. A mass is moving in simple harmonic motion described by

$$x = 3e^{-t/4} \cos(3t - 15°) \text{ ft}, \qquad t \geq 0$$

where t is in seconds. Find the last time t for which $|x| = 1$ foot.

20. Johannes Kepler (1571–1630) developed the first mathematical description of the motion of a planet around the sun. The orbit is an ellipse with the sun at one focus. Kepler found it useful to describe the motion of a planet around the orbit in terms of motion at constant speed around a circle circumscribing the elliptical orbit and tangent to it at perihelion (closest approach to the sun) and at aphelion (farthest from the sun). If the planet is at the point P in Figure 5.10, the angle ACQ is called the eccentric anomaly E. The angle M (not shown), called the mean anomaly, is the angle that the planet would make (with AC) if it traveled around the circular orbit at constant rate with the same period as it has in the elliptical orbit. Kepler derived an equation relating E and M:

$$M = E - e \sin E$$

where e is the eccentricity of the orbit. This equation makes it simple to find the position of the planet in terms of the time after perihelion passage, to which M is directly proportional.

The planet Jupiter has a period of 11.8622 years (a year has 365.2425 days) and has an orbit with eccentricity $e = 0.04844$. Find the eccentric anomaly of Jupiter for a time 1505 days after passage through perihelion.

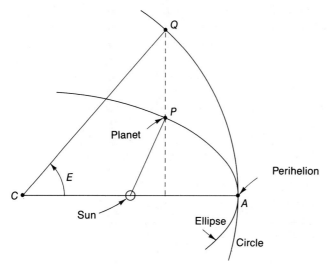

FIGURE 5.10

21. A 4-foot length of tubing is to be bent into the arc of a circle so that the ends are 3 feet apart. What is the radius of the arc?

22. A chain or perfectly flexible cable supported at its ends takes a shape called a *catenary*. If coordinates are chosen such that $(x, y) = (0, a)$ at the low point on this curve, then the equation of the catenary is

$$y = \left(\frac{a}{2}\right)(e^{x/a} + e^{-x/a}) = a \cosh\left(\frac{x}{a}\right)$$

A cable is suspended from supports at equal height and 20 feet apart. The middle of the cable sags exactly 10 feet below the level of the supports. What is the value of a?

23. Insertion sort and Quicksort are two of the many methods of ordering a set of numbers. Insertion sort is a slow method with the property that doubling the number N of items to be sorted approximately quadruples the sorting time on a computer. Quicksort has a sorting time for random data proportional to the factor $N \log N$, which grows with N at a much slower rate than for insertion sort. However, Quicksort can become as slow as insertion sort if the data are already sorted unless the program is modified.

D. E. Knuth [1973] analyzed the running times for insertion sort and the median-of-three version of Quicksort for a hypothetical computer and found the following dependencies:

$$2N^2 + 9N \text{ units of time} \qquad \text{Insertion sort}$$
$$11.60N \ln N + 1.89N \quad \text{units} \qquad \text{Quicksort}$$

Find the "crossover point," or the value of N below which insertion sort is faster than median-of-three Quicksort on Knuth's machine. (*Hint:* Replace the discrete variable N by a continuous variable x. Then test your answer after rounding x up and down to the nearest integers.)

24. Find all the roots of the following polynomial equations:

a. $x^4 + 6x^3 + 18x^2 + 30x + 25 = 0$
b. $4x^5 - 8x^4 + 19x^3 - 38x^2 - 5x + 10 = 0$
c. $25x^4 + 30x^3 + 18x^2 + 6x + 1 = 0$
d. $12x^5 - 20x^4 + 327x^3 - 554x^2 - 609x + 814 = 0$
e. $8x^4 - 24x^3 + x - 3 = 0$
f. $x^4 + x^3 + x^2 + x + 1 = 0$
g. $x^5 + x^4 + x^3 + x^2 + x + 1 = 0$

25. The dimensions of a rectangular metal enclosure are 6 in. by 8 in. by 12 in. What are the dimensions of another enclosure that has exactly twice the volume if each dimension is increased by the same amount?

Problems

Project Problem 5.1

Another two-point method of solving nonlinear equations is known as *regula falsi*. In the secant method, the oldest point (which is arbitrarily chosen for the two starting points) is replaced by the new point in seeking the approximate root, and there is no constraint on the new point's location. In the regula falsi method, the two starting points must bracket the root such that $f(x_a) \cdot f(x_b) < 0$ if x_a and x_b are the abscissas of the two starting points. When x_c is found, it replaces either x_a or x_b so that $f(x_a) \cdot f(x_b) < 0$ remains true after the replacement.

Using a geometrical construction, derive the equation

$$x_c = x_a - f(x_a)\frac{x_b - x_a}{f(x_b) - f(x_a)}$$

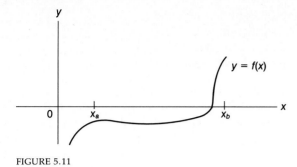

FIGURE 5.11

or the equivalent equation in which x_a and x_b are interchanged on the right-hand side.

Project Problem 5.2

Show geometrically that the solution of $f(x) = 0$ by the method of regula falsi is slow for the case of a function $f(x)$ whose graph is shown in Figure 5.11 and where x_a and x_b are chosen as shown. Show how Newton's method would behave for the choices of x_a and x_b as starting values.

Project Problem 5.3

A natural-gas reservoir can be modeled by assuming that it has a water drive, which tends to sustain the gas pressure and causes gas entrapment as the water invades the reservoir when gas is withdrawn [Hultquist, 1978], [Collier, 1979]. If the rate at which gas is withdrawn is assumed to be proportional to the square of the gas pressure, then a relatively simple equation governs the relationship of gas pressure and reservoir volume (unencroached by water):

$$x = x^{-(1+F)} \left[\frac{y - 1/(1 + k)}{1 - 1/(1 + k)} \right]^{b}$$

where
$b = k(1 + F)/(1 + k)$
F = entrapment constant, dimensionless
$x = V/V_0$, normalized volume, dimensionless
$y = P/P_0$, normalized pressure, dimensionless
$k = r_0RT/CP_0^2$, dimensionless
r_0 = initial production rate, moles per mole per year
P_0 = initial reservoir pressure, psia

V_0 = initial reservoir volume, cubic feet
R = universal gas constant = $19.33\,\text{ft}^3\,\text{psia}/(\text{mol}\,°\text{K})$
T = reservoir temperature, degrees Kelvin
C = Schilthuis water drive constant, $\text{ft}^3/(\text{yr psia mol})$

Typical curves of y versus x are shown in Figure 5.12.

A reservoir will be useful until the pressure drops to a value that makes production uneconomic or until the volume drops to an uneconomical value (many wells become "watered out").

Let x_{co} and y_{co} be the (normalized) volume and pressure below which production is uneconomic. Write a subroutine called FIND that will determine for all values of k_0 and F the values (x_c, y_c), where the solution curve intersects one of the cutoff lines $x = x_{co}$ or $y = y_{co}$.

Use the following data:

$F = 0, 0.5, 1.0$

$C = 10^{-3}, 10^{-5}, 10^{-7}\,\text{ft}^3/(\text{yr psia lbm mol})$

$r_0 = 0.05, 0.10, 0.20\,\text{mol/mol}$

$T = 550\,°\text{K}$

$x_{co} = y_{co} = 0.15$

$P_0 = 10^4\,\text{psia}$

Be sure that your program can distinguish correctly between solutions that terminate on the y_{co} line and the x_{co} line.

FIGURE 5.12 Normalized pressure-volume curves for water drive reservoir under P^2 rate production.

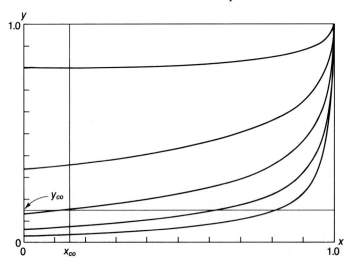

Project Problem 5.4

A chemical batch reactor produces a quantity Q of finished chemical product according to the formula

$$Q = Q_0 (1 - e^{-kt})$$

where Q_0 is the limiting amount, t is the time from start to whenever the process is stopped, and k is the reaction rate. At some time t the process is stopped and the reactor is opened, emptied, and recharged with raw materials. The time for emptying, cleaning, and refilling is t_C. Thus the average amount of product produced by the reactor in time $t + t_C$ is

$$A = \frac{Q}{t + t_C}$$

This can be maximized by setting $dA/dt = 0$ and solving for the value of t.

1. Find dA/dt and set it to 0 in order to derive an equation involving t that can be solved to find the optimum running time for each reactor cycle.
2. Write a program to solve the equation in part (1). This program should either produce graphical output or the data for plotting curves of optimum t versus reaction rate k. The desired range of k is from $k = 0.1$ to $k = 5.0\,\mathrm{h}^{-1}$. Show curves for values of $t_C = 0.2, 0.5, 1.0, 2.0\,\mathrm{h}$.

Project Problem 5.5

The formula for a time payment P, which in n periods (months, typically) will repay an amount A borrowed at a rate i of interest is given by

$$P = \frac{Ai(1 + i)^n}{(1 + i)^n - 1}$$

(If the interest rate is an annual rate, such as 18%, the monthly rate is $\frac{18}{12} = 1.5\%$. The interest rate i as used in the formula is always a decimal fraction, so that the factor $1 + i$ for monthly payments at 18% annual interest is 1.015.)

1. Derive the formula using the idea that at the end of the first month you owe $A(1 + i)$ and after the payment you owe only $A(1 + i) - P$. This is the new principal, so at the end of the second month you owe $[A(1 + i) - P](1 + i)$, and so on. At the end of n months, after making the last payment, you owe nothing, so you can set the expression started above equal to 0 and solve for P. (Bankers never deal in amounts of money less than $0.01, so the payments seldom come out paying the loan exactly. Hence the last payment has to be adjusted slightly to take care of the round-off error.)
2. Develop an algorithm for calculating the rate of interest, given A, P, and

n. In choosing a starting value for *i* for the iteration, note that $Ai < P$, or else the loan can never be repaid.

3. Check out the algorithm for various values of the parameters. Choose values of A, i, and n to calculate P; then use A, n, and P to calculate i.

4. Modify the formula to fit the following set of facts: A prize contest gives the winner a choice of \$90,000 cash now or annual payments of \$10,000 for 25 years. If we assume that the contest manager knows how to use interest formulas and invests the \$80,000 left after the initial payment, what is the rate of interest that the manager is expecting to get?

   Hint: The sequence goes like this:

$$A - P = \text{amount manager has after initial}$$
$$\text{payment}$$
$$(A - P)(1 + i) = \text{amount manager has at the end}$$
$$\text{of a year}$$
$$(A - P)(1 + i) - P = \text{amount manager has after second}$$
$$\text{payment}$$
$$(A - P)(1 + i)^2 - P(1 + i) = \text{amount manager has at the end}$$
$$\text{of the second year}$$
$$(A - P)(1 + i)^2 - P(1 + i) - P = \text{amount manager has after the}$$
$$\text{third payment}$$

   This is continued until, at the end of the 25th payment, the expression is equal to zero.

5. Determine the annual interest rate assumed by the contest manager.

Project Problem 5.6

Write an interactive polynomial equation solver incorporating the procedure BAIRSTOW. The driver program should ask the user for the degree of the polynomial and the set of coefficients.

Modify BAIRSTOW in some or all of the following ways:

1. Move the output statements to the driver program or to another procedure. Return the root pairs from BAIRSTOW in a 2 × 2 array. Store all the roots in an array in the driver program until BAIRSTOW has finished.

2. Remove the HALT statements from BAIRSTOW and provide an integer flag to return a number to designate the nature of the input error or the end of the root-finding process. Provide appropriate error messages to be printed from the driver program and allow for correction of input data by the user.

3. Allow the user to specify the epsilon(s) for the convergence test.

4. Devise a test to determine whether BAIRSTOW is converging slowly or is wandering aimlessly around the *z* plane if the convergence test is not

met in 300 iterations. If it appears to be a case of slow convergence, let the user decide among several choices, such as:

a. Relax the convergence test epsilon(s).

b. Print r and s for a few iterations for user observation.

c. Stop the program.

If it appears to be a case of "hanging" or "thrashing," let the user decide whether to stop or to choose new values of r and s.

5. Allow the user to enter estimates of roots. Use these estimates to create initial estimates of r and s.

Create several test-case polynomials with known solutions to test the program.

Project Problem 5.7

Write an interactive equation solver using the procedure ZEROIN. Allow the user to provide a function $f(x)$ for which the real roots of $f(x) = 0$ are desired. If Turbo PASCAL is used, the function can be included by means of a {$I filename} statement.

Provide for entry by the user of A and B (values of x that bracket the desired root) and of the epsilon for terminating the program. Also provide for appropriate error messages, such as for the case where $f(A) \times f(B) > 0$.

Create several suitable test problems and solve them to make sure that the program works properly.

Project Problem 5.8

Figure 5.13 shows a full-wave rectifier circuit with a capacitor filter. T is a transformer with a center-tapped secondary, D is a diode rectifier, and C is the filter capacitor.

A mathematical model of the output voltage V is obtained by sketching the full-wave, rectified wave form in the absence of C. See Figure 5.14. With C in

FIGURE 5.13

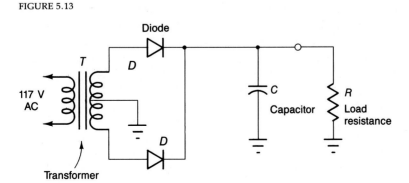

Transformer

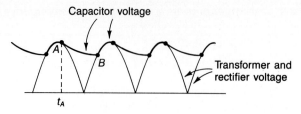

FIGURE 5.14

place, the capacitor voltage (voltage V) follows the *leading edge* of the transformer voltage, but on the *trailing edge* the voltage can fall only as rapidly as the load resistance R will allow. When the rate of decrease of the transformer voltage exceeds the rate of decrease permitted by C and R, the voltage will follow an exponential decay curve and the diode will cease to conduct. This will happen at point A in Figure 5.14, where the decay curve and the sinusoid have the same slope.

The decay curve has the equation

$$V = V_A e^{-t/RC} \text{ V}$$

where V_A is the voltage at point A and t is measured beginning at point A.

At point B, the transformer voltage rises to the capacitor voltage and the diode again conducts.

1. Find the value at which the decay curve has the same slope as the sinusoid. This does not require iteration.
2. Find the time at which the diode again begins to conduct (point B). This does require iteration. (These times should be calculated with respect to the time at which the transformer voltage reaches a maximum.) Use the following data:

$$V = 100 \text{ V}$$
$$R = 10, 100, 1000 \, \Omega$$
$$C = 10, 100, 1000 \, \mu\text{F}$$

in all combinations

3. Calculate the voltage drop from $V_{\max} = 100$ V to V_B for each case.

Listings

LISTING OF ZEROIN

```
FUNCTION ZEROIN(    AX:   REAL;
                    BX:   REAL;
                    TOL:  REAL   ):   REAL;
```

```
VAR    A, B, C, D, E, EPS, FA, FB, FC:     REAL;
            TOLI, XM, P, Q, R, S:     REAL;
                    LOOP:     INTEGER;
```

{ZEROIN finds a root of F(X) = 0 in the interval AX, BX.

Input: AX   Left endpoint of initial interval
          BX   Right endpoint of initial interval
           F   Function which evaluates F(X) for any X in
                  the interval
          TOL   Desired length of the interval of
                  uncertainty of the final result.
                  TOL => 0.0

Output: ZEROIN   Abscissa approximating a zero of F in
                     [AX, BX]

It is assumed that F(AX) and F(BX) have opposite signs
without a check.   ZEROIN returns a zero X in the interval
[AX, BX] to within a tolerance of 4*EPS*ABS(X) + TOL where
EPS is the machine epsilon.

This function is a modified translation by Craig Schons of
the ALGOL 60 procedure ZERO given by Richard Brent,
Algorithms for Minimization without Derivatives, Prentice-
Hall, 1973.}


```
FUNCTION SIGN(X, Y: REAL): REAL;
  BEGIN
    IF Y>=0.0 THEN SIGN := ABS(X)
    ELSE SIGN := -ABS(X)
  END;

BEGIN

  LOOP := 0;                 {Compute EPS, the machine epsilon}

  EPS := 1.0;
  REPEAT
    EPS := EPS/2.0;
    TOLI := 1.0 + EPS;
  UNTIL TOLI = 1.0;

  A := AX;                              {Initialization}
  B := BX;
```

```
  FA := F(A);
  FB := F(B);

  C := A;                               {Begin step}
  FC := FA;
  D := B - A;
  E := D;

REPEAT
  IF ABS(FC) < ABS(FB) THEN
    BEGIN
      A := B;
      B := C;
      C := A;
      FA := FB;
      FB := FC;
      FC := FA;
    END;

TOLI := 2.0*EPS*ABS(B) + 0.5*TOL;   {Setup for convergence}
XM := 0.5*(C - B);                                    {test}
  IF (ABS(XM) > TOLI) AND (FB <> 0.0) THEN
    BEGIN
                {Bisection necessary?}

      IF (ABS(E) >= TOLI) AND (ABS(FA) > ABS(FB)) THEN
        BEGIN
          IF A = C THEN   {Quadratic interp. possible?}
            BEGIN

              S := FB/FA;              {Inverse quadratic}
              P := 2.0*XM*S;          {interpolation}
              Q := 1.0 - S
            END
          ELSE
            BEGIN

              Q := FA/FC;              {Linear interpolation}
              R := FB/FC;
              S := FB/FA;
              P := S*(2.0*XM*Q*(Q - R)
                  - (B - A)*(R - 1.0));
              Q := (Q - 1.0)*(R - 1.0)*(S - 1.0)
            END;

          IF P>0.0 THEN Q := -Q;        {Adjust signs}
          P := ABS(P);
                                {Interpolation acceptable?}

          IF ((2.0*P) < (3.0*XM*Q - ABS(TOLI*Q))) AND
            (P < ABS(0.5*E*Q)) THEN
```

```
            BEGIN
               E := D;
               D := P/Q
            END
          ELSE
        BEGIN                            {Bisection}
          D := XM;
          E := D
        END
      END
    ELSE BEGIN
      D := XM;
      E := D
    END;
                                         {Complete the step}

    A := B;
    FA := FB;
    IF ABS(D)>TOLI THEN
      B := B + D
    ELSE
      B := B + SIGN(TOLI,XM);
    FB := F(B);
    IF FB*FC/ABS(FC)>0.0 THEN
      BEGIN
        C := A;
        FC := FA;
        D := B - A;
        E := D
      END

    END                                  {Of main IF}

    ELSE LOOP := -1;
  END

  UNTIL LOOP=-1;

  ZEROIN := B;

END;
```

LISTING OF BAIRSTOW

```
PROCEDURE BAIRSTOW(   A: VECTOR;
                      N: INTEGER);

{Procedure BAIRSTOW attempts to find all roots of a
polynomial equation with real coefficients.
```

The calling program must define the data type VECTOR:

```
VECTOR = ARRAY[-2..M]
```

where M is at least as great as 1 + degree of the largest
polynomial equation expected to be solved.

On input, vector A contains the coefficients such that
A[0] is the coefficient of the highest-degree term in the
equation. N is the degree of the equation to be solved.
Procedure BAIRSTOW provides its own output. }

```
VAR                B,C:  VECTOR;
                 I,J,K:  INTEGER;
        RROOT1,RROOT2:  REAL;
                 IROOT:  REAL;
             R,S,RAD:  REAL;
     DENOM,RNUM,SNUM:  REAL;
       DELTAR,DELTAS:  REAL;
                 TEST:  REAL;
                 FLAG:  BOOLEAN;

LABEL 1,2;

PROCEDURE EQROOTS;
  BEGIN
    RROOT1 := -R/2.0;
    WRITELN;  WRITELN(RROOT1);
    WRITELN;  WRITELN(RROOT1);
  END;

PROCEDURE IMAGROOTS;
  BEGIN
    RROOT1 := -R/2.0;
    IROOT := SQRT(ABS(RAD))/2.0;
    WRITELN;  WRITELN(RROOT1,'  +i',IROOT);
    WRITELN;  WRITELN(RROOT1,'  -i',IROOT);
  END;

PROCEDURE REALROOTS;
  BEGIN
    RAD := SQRT(RAD);
    IF R < 0 THEN RROOT1 := (-R + RAD)/2.0
      ELSE RROOT1 := (-R - RAD)/2.0;
    RROOT2 := S/RROOT1;
    WRITELN;  WRITELN(RROOT1);
    WRITELN;  WRITELN(RROOT2);
  END;
```

```
PROCEDURE LINEAR;
  BEGIN
    RROOT1 := -B[1]/B[0];
    WRITELN; WRITELN(RROOT1);
  END;

PROCEDURE QUADRATIC;
  BEGIN
    RAD := R*R - 4.0*S;
    IF RAD < 0 THEN IMAGROOTS ELSE
    IF RAD = 0 THEN EQROOTS ELSE
    IF RAD > 0 THEN REALROOTS;
  END;

BEGIN {Bairstow}

  A[-2] := 0;  A[-1] := 0;
  B[-2] := 0;  B[-1] := 0;
  C[-2] := 0;  C[-1] := 0;

  IF N < 1 THEN
    BEGIN
      WRITELN('N < 1. Halt.');
      HALT
    END;

  IF N = 1 THEN             {Linear equation}
    BEGIN
      B[0] := A[0]; B[1] := A[1];
      LINEAR;
      HALT;
    END;

  IF N = 2 THEN            {Quadratic equation}
    BEGIN
      IF A[0] <> 0 THEN
        BEGIN
          R := A[1]/A[0];
          S := A[2]/A[0];
          QUADRATIC;
          HALT;
        END
      ELSE
        BEGIN
          B[0] := A[1]; B[1] := A[2];
          LINEAR;
          HALT;
        END;
    END;
```

```
1:   R := 1.0;
     S := 1.0;
     FLAG := FALSE;
     K := 0;
2:   J := 0;

     REPEAT                        {General case of N > 2}
       BEGIN
         FOR I := 0 TO N DO
           BEGIN
             B[I] := A[I] - R*B[I-1] - S*B[I-2];
             C[I] := B[I] - R*C[I-1] - S*B[I-2];
           END;
         DENOM := C[N-2]*C[N-2] - C[N-3]*(C[N-1] - B[N-1]);
           IF ABS(DENOM) < 1.0E-20 THEN
             BEGIN
               R := R + 1.9;        {Try again with different}
               S := S + 2.30123;    {R and S and hope that    }
               GOTO 2;              {DENOM <> 0               }
             END;

         RNUM := B[N-1]*C[N-2] - B[N]*C[N-3];
         SNUM := B[N]*C[N-2] - B[N-1]*(C[N-1] - B[N-1]);
         DELTAR := RNUM/DENOM;
         DELTAS := SNUM/DENOM;
         R := R + DELTAR;
         S := S + DELTAS;            {Recalculate R and S}
         J := J + 1;
         TEST := B[N]*B[N] + B[N-1]*B[N-1];
       END;

     IF TEST < 1.0E-6 THEN FLAG := TRUE;
     IF (ABS(DELTAR) < 1.0E-6) AND (ABS(DELTAS) < 1.0E-6)
        THEN FLAG := TRUE;
     UNTIL (J > 100) OR (FLAG = TRUE);

     IF J > 100 THEN
       BEGIN                        {Try a new R and S}
         R := R + 5.0;
         S := S + 5.0;
         K := K + 1;
         IF K >= 12 THEN
           BEGIN
             WRITELN('No Convergence');
             HALT;
           END
         ELSE  GOTO 2;
       END;
```

```
    QUADRATIC;                {Two more roots found}
        N := N - 2;
        IF N = 1 THEN
          BEGIN
            LINEAR;           {All done; N odd}
            HALT;
          END;
        IF N = 2 THEN
          BEGIN
            R := B[1]/B[0];
            S := B[2]/B[0];
            QUADRATIC;        {All done; N even}
            HALT;
          END;
        FOR I := 0 TO N DO A[I] := B[I];    {Deflate the}
        GOTO 1;                             {polynomial }

END;
```

6

Fitting Curves to Data

A common problem in science and engineering is the representation of a function when only a few points on its graph are accurately known. As an example, Table 6.1 shows values of the drag coefficient C_D for a nonrotating sphere traveling with Mach number M through a fluid. The Mach number (named for Ernst Mach, 1838–1916, an Austrian physicist and philosopher) is the ratio of the speed of the sphere with respect to the fluid to the speed of sound in the fluid. The value of C_D is needed to calculate the drag force F_D on the sphere given by

$$F_D = \frac{A \rho C_D V^2}{2}$$

where A is the cross-sectional area of the sphere, ρ is the fluid density, and V is the speed of the sphere in the fluid.

Our intuition tells us that we should be able to pass a smooth curve through the points from the table in order to obtain a graphical representation of the function. We also can use the function defining the curve to calculate values of the drag coefficient for values of M that lie between the given points. This ability is essential to many applications, such as computer simulation of physical systems.

Figure 6.1 shows how the curve looks when we use a tenth-degree polynomial passing through the 11 points. Most of us are likely to feel that the "wiggles" near the ends of the curve are not satisfactory, and those persons would be much happier with the curve shown in Figure 6.2. The latter curve is produced by a set of 10 cubic polynomial pieces, known as cubic splines, which join at the given points and which match in slope and curvature at those points. In this chapter we shall see how to produce both spline and polynomial inter-

Table 6.1   Drag Coefficient C_D versus Mach Number

Mach	C_D
0.0	0.50
0.4	0.52
0.8	0.66
1.2	0.93
1.6	1.03
2.0	1.01
2.4	0.99
2.8	0.97
3.2	0.95
3.6	0.93
4.0	0.92

polators as well as how to find curves that represent sets of "noisy" data containing measurement errors.

A philosophical question still remains: How do we know that the function does not, in fact, wiggle or have discontinuities or sharp corners on its graph between the data points? We do not, of course, but the presumption is that the investigator who produced the points from measurements has studied and measured the phenomenon with care. If such anomalies do exist, then the investigator is obliged to report them. The structure of science and engineering rests on the integrity of the scientists and engineers who have studied the phenomena of nature and have documented their results carefully.

FIGURE 6.1   Polynomial interpolation of drag coefficient.

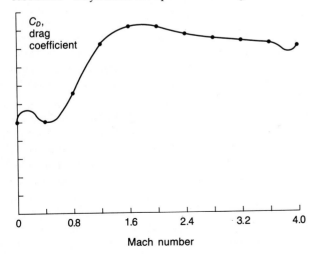

Mach number

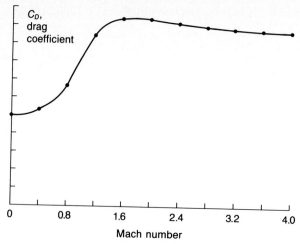

FIGURE 6.2   Spline interpolation of drag coefficient.

6.1   Preliminaries

In this chapter we will look at the solution of several different, but closely related, problems. The first of these is the classical interpolation problem. Here we have a table of values of some function, such as the sine or the logarithm, usually given by values at a set of equally spaced points, from which we wish to determine the functional value at some point between the tabulated values. This problem is not of as much interest as it was a generation ago, but the solution of it brought considerable understanding to some of the problems of today. For this reason we will spend a little time on the facets of interpolation.

The second problem comes about because a computer must compute functions such as logarithms, sines, or exponentials. The only capability it has, other than logical, is the ability to do simple arithmetic. Hence it is limited to computing the simplest of functions: polynomials and rational functions, which are quotients of polynomials. The problem is then to devise suitable polynomial or rational approximations to the desired functions. This problem will be dealt with in Chapter 9.

A third problem is that of determining a best-fitting (in some sense) curve that passes through a set of points that might best be described as experimentally determined, or *noisy*, data. The function to be used to represent the data may be chosen on the basis of underlying theory (e.g., radioactive decay gives rise to exponential curves with negative exponents if a single species is involved), or it may be chosen expediently as a polynomial or rational function simply to show the trend of the data.

Related to these is the numerical integration problem, which is that of replacing the integrand of a difficult-to-handle definite integral by an interpolating

polynomial or a set of splines that can be integrated easily. This is such an important and specialized case that we will defer it to Chapter 7 and spend our time
here learning about the underlying interpolation processes.

6.2 Functions and Continuity

A function is a mapping from one set of points (the domain) to another set (the
range), in which there is only one point in the range set for each point in the
domain set. In our ordinary graphical representation, we think of the range set
as lying along the y axis (ordinates) and the domain set as lying along the x axis
(abscissas). Sometimes we write this as $f : R \to R$ to indicate that the function is
f and we are mapping from the real numbers to the real numbers. Less formally,
$y = f(x)$. In the case of two variables, this might be written as $g : R^2 \to R$, or z
$= g(x, y)$.

In most of the cases in this book the functions involved are continuous. In
all cases they are single-valued. This means that a vertical line can intersect the
graph of the function at most once, or that there is only one number in the range
set for each number in the domain set. The student is assumed to have a good
grasp of the meaning of continuity, at least in the geometric sense. We say that f
is continuous at $x = a$ if and only if (iff)

$$\lim_{x \to a} f(x) = f(a)$$

More formally, this means

Given $0 < \epsilon$, arbitrarily small, there exists $\delta > 0$ such that $|f(x) - f(a)| < \epsilon$
whenever $|x - a| < \delta$.

Note that either definition requires that $\lim f(x)$ exist, that $f(a)$ be defined, and
that the two be equal. Figure 6.3 shows the graph of a function with a jump-
type discontinuity at $x = a$ and an infinite discontinuity at $x = b$. At $x = a$ we
have two different limits of $f(x)$, depending upon whether x is approaching a
from above or below. (In such a situation $f(a)$ may be undefined, equal to one or
the other limit, or equal to something else.) At $x = b$ the value of $f(x)$ is undefined
and the limit does not exist.

6.3 Interpolation

Suppose we have a set of n points $\{x_k\}$ and the corresponding set $\{f(x_k)\}$ of functional values for $k = 1, 2, \ldots$, where $1 \leqslant k \leqslant n$. One way of approximating
$f(x)$ is to replace $f(x)$ by a set of straight-line segments that join $(x_k, f(x_k))$ and
$(x_{k+1}, f(x_{k+1}))$ for all values of k, as shown in Figure 6.4. This produces a set of
linear functions $L_k(x)$, which join at the points so that the collection of these

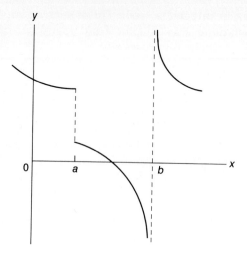

FIGURE 6.3   Two types of discontinuities.

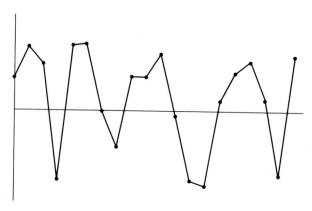

FIGURE 6.4

forms a continuous function over the desired domain. The original function from which the points were taken may have had a smooth curve over the domain. However, the set of linear segments is not smooth, although it is continuous. It has a discontinuous derivative at each of the "knots," or points where the segments join because the straight-line segments have, in general, different slopes. It is the simplest example of a spline curve. Most splines that are used are composed of segments of higher-degree polynomials, so that there are not sharp corners on the approximating function graph.

It is left as an exercise to show by geometrical construction that each straight-line segment has the equation

$$y = \alpha f(x_{k+1}) + (1 - \alpha)f(x_k)$$

where α is a local variable, which ranges from 0 to 1 over the interval (x_k, x_{k+1}). Here, α can be thought of as the fractional distance from x_k to x_{k+1}, or

$$\alpha = \frac{x - x_k}{x_{k+1} - x_k}$$

The process of linear interpolation entails the replacement of a small piece of the function by a straight-line approximation. It is of limited use, mostly for finding values of relatively low accuracy from tables. We are more interested in creating interpolation formulas that allow us to pass smooth curves through the given points. These will yield more accurate values between the given (base) points. This requires polynomials of higher degree than first degree and splines of higher order than first order.

6.3.1 Lagrange Interpolation Polynomials

Lagrange interpolation polynomials are constructed in the following way. Consider a set of n points $\{x_i\}$ and a corresponding set of functional values $\{f(x_i)\}$. First we construct a set of Lagrange polynomials of the following form:

$$L_j(x) = \frac{(x - x_1)(x - x_2)\cdots(x - x_{j-1})(x - x_{j+1})\cdots(x - x_n)}{(x_j - x_1)\cdots(x_j - x_{j-1})(x_j - x_{j+1})\cdots(x_j - x_n)}$$

This can also be written as

$$L_j(x) = \frac{\prod\limits_{i \neq j}(x - x_i)}{\prod\limits_{i \neq j}(x_i - x_j)}$$

The polynomials $L_j(x)$ have the following properties:

1. $L_j(x_i) = 0,$ for $i \neq j$.
2. $L_j(x_j) = 1$.
3. $L_j(x)$ is a polynomial of degree $n - 1$ if there are n points in the set $\{x_i\}$.

You should verify these properties with the aid of a simple example.

Now suppose we have n base points x_j and n corresponding functional values $f(x_j)$. We wish to construct a polynomial through the set of points $\{x_j, f(x_j)\}$. We can do this by writing a polynomial formula

$$p_{n-1}(x) = \sum_{j=1}^{n} L_j(x)f(x_j)$$

It remains to be shown that this is, indeed, an $(n - 1)$st-degree interpolating polynomial.

It is clear that the interpolating polynomial passes through the points. If x_k is any base point,

$$p_{n-1}(x_k) = \sum_{j=1}^{n} L_j(x_k)f(x_j) = f(x_k)$$

because for all $j \neq k$, $L_j(x_k) = 0$, but for $j = k$, $L_k(x_k) = 1$. Thus $p_{n-1}(x_k)$ reduces to $f(x_k)$. Clearly the representation is an $(n - 1)$st-degree polynomial because each of the terms in the sum is an $(n - 1)$st-degree polynomial.

Example 6.1

Using Lagrange polynomials, construct an interpolation polynomial passing through the points (1, 1), (2, 2), (3, 5), (4, 2) using Lagrange polynomials.

$$L_1 = \frac{(x-2)(x-3)(x-4)}{(1-2)(1-3)(1-4)} = \frac{-(x-2)(x-3)(x-4)}{6}$$

$$L_2 = \frac{(x-1)(x-3)(x-4)}{(2-1)(2-3)(2-4)} = \frac{(x-1)(x-3)(x-4)}{2}$$

$$L_3 = \frac{(x-1)(x-2)(x-4)}{(3-1)(3-2)(3-4)} = \frac{-(x-1)(x-2)(x-4)}{2}$$

$$L_4 = \frac{(x-1)(x-2)(x-3)}{(4-1)(4-2)(4-3)} = \frac{(x-1)(x-2)(x-3)}{6}$$

$$p_3(x) = -\left[\frac{(x-2)(x-3)(x-4)}{6}\right] \cdot 1 + \left[\frac{(x-1)(x-3)(x-4)}{2}\right] \cdot 2$$
$$+ \left[\frac{-(x-1)(x-2)(x-4)}{2}\right] \cdot 5 + \left[\frac{(x-1)(x-2)(x-3)}{6}\right] \cdot 2$$

This is an easy method to understand but it is not well suited for computing. If we reduce the inefficient Lagrange form to a properly nested form suitable for computation, we have:

$$p_3(x) = (((-1.333\ldots)x + 9)x - 16.666\ldots)x + 10$$

(which yields the polynomial value with only 3 multiplications and 3 additions). This reduction requires much algebra, either in the machine or on paper. Moreover, the reduction also may introduce ill-conditioning. For this reason we give only a straightforward PASCAL function called LAGRANGE for evaluating Lagrange polynomials.

The call to LAGRANGE is the standard function call in which the name of the function appears on the right-hand side in a replacement statement. The parameters in the call are

LAGRANGE (X, Y, N, V)

where X is the vector of x's, Y is the vector of y's (the x's and y's in X and Y must be in the same order), N is the number of points in the set of points (x_i, y_i), and V is the value of x for which the polynomial is to be evaluated. N is type INTEGER, V is type REAL, and X and Y are type VECTOR, where the type must be declared in the calling program by a statement

```
TYPE    VECTOR = ARRAY[1..M] OF REAL;
```

The value of M in the declaration must, of course, be at least as large as the value of N. (If memory is of no great concern, it is often helpful to declare vectors to be of some large size that will be sufficient for all cases, such as 25 or 50. It is unlikely that one would ever use polynomials of such high degree.)   □

6.3.2 Difference Methods

In order to be able to understand several powerful methods of interpolation later in the chapter, we need to look at what are called *differences*. We introduce the following notation:

$$\Delta f(x) = f(x + h) - f(x)$$

is the forward difference of $f(x)$, where $h > 0$ is some (generally small) distance along the x axis. We will find it convenient to think of Δ as a differencing operator applied to a function, in some senses analogous to the derivative operator. (The divided difference $\Delta f(x)/h$ is actually more nearly the discrete counterpart of the derivative; Δ is more closely related to the differential.) We can define higher-order differences, such as

$$\Delta^2 f(x) = \Delta(\Delta f(x)) = \Delta(f(x + h) - f(x)) = \Delta f(x + h) - \Delta f(x)$$
$$= f(x + 2h) - 2f(x + h) + f(x)$$

In general,

$$\Delta^n f(x) = \Delta(\Delta^{n-1} f(x))$$
$$= f(x + nh) - nf(x + (n - 1)h) + \cdots + (-1)^n f(x)$$
$$= \sum_{k=0}^{n} (-1)^k \binom{n}{k} f(x + (n - k)h)$$

where $\binom{n}{k}$ is the usual symbol for the binomial coefficient.

In addition, we can define backward differences

$$f(x) = f(x) - f(x - h)$$
$$\nabla^2 f(x) = f(x) - 2f(x - h) + f(x - 2h)$$

.

.

.

$$\nabla^n f(x) = f(x) - nf(x - h) + \cdots + (-1)^n f(x - nh)$$
$$= \sum_{k=0}^{n} (-1)^k \binom{n}{k} f(x - kh)$$

Table 6.2   Table of Forward (Δ) and Backward (∇) Differences

Table 6.2 shows a set of values of x_i and $f(x_i)$ and the arrangement of the differences in a table. Here the subscript of x is attached to f to save space. That is, $\Delta f(x_2)$ is replaced by Δf_2. Note that we write the first and other odd differences on lines between the functional values and we write the even differences on the same lines as the functional values.

Table 6.3 shows a part of an actual table of sines tabulated every 1° together with the differences through the fourth. The differences are found simply by subtraction: $0.01403 = 0.60182 - 0.58779$ and $-0.00019 = 0.01384 - 0.01403$. Differences are generally done in the direction of the increasing independent variable, the angle in this case. If we think of 40° as x_0 and 41° as x_1, then $\Delta f_0 =$

Table 6.3   Sine Function at Integral Degrees

x (deg)	Sine	Δ, ∇	Δ^2, ∇^2	Δ^3, ∇^3	Δ^4, ∇^4
36	0.58779				
		0.01403			
37	0.60182		-0.00019		
		0.01384		0.00001	
38	0.61566		-0.00018		-0.00002
		0.01366		-0.00001	
39	0.62932		-0.00019		0.00000
		0.01347		-0.00001	
40	0.64279		-0.00020		0.00001
		0.01327		0.00000	
41	0.65606		-0.00020		0.00000
		0.01307		0.00000	
42	0.66913		-0.00020		
		0.01287			
43	0.68200				

$f_1 - f_0 = 0.01327$. However, $\nabla f_1 = f_1 - f_0 = \Delta f_0$. This means that in difference tables a line running downward diagonally from f_1 in the table will pass through all the forward differences with subscript 1. Likewise, a line running diagonally upward from f_1 will pass through backward differences with subscript 1 (see Table 6.2).

Newton Difference Formulas for Interpolation Analogous to Taylor series formulas are the many difference formulas that can be used for interpolation. Taylor series are not useful for this purpose because they require (at one point only) the value of the function and the values of several of its derivatives. Only the values of the function at several points are available in a typical interpolation problem. In the difference formulas, the derivatives are replaced by differences of various orders, and these, of course, involve values of the function at several points. One of the most common of these formulas is Newton's forward formula (NFF):

$$f(x_0 + \alpha h) = f(x_0) + \alpha \Delta f(x_0) + \frac{\alpha(\alpha - 1)}{2!} \Delta^2 f(x_0) + \cdots$$

$$+ \frac{\alpha(\alpha - 1)\cdots(\alpha - n + 1)}{n!} \Delta^n f(x_0) + \frac{\alpha(\alpha - 1)\cdots(\alpha - n)}{(n + 1)!} f^{(n+1)}(\xi) h^{n+1}$$

where h is the common distance between successive x's, α is the fractional distance from x_0 to the point of interpolation in terms of the length h, $f^{(n+1)}$ designates the $(n + 1)$st derivative of f, and ξ is somewhere in the interval bounded at the extremes by x_0, x_n, or $x = x_0 + \alpha h$ if x is outside the interval (x_0, x_n).

This formula can be used, for example, to estimate the value of $\sin 37°45'$. If we select x_0 as 37, then $\alpha = \frac{45}{60} = 0.75$.

$$f(37°45') = \sin 37° + (0.75)\Delta f(37) + \frac{(0.75)(-0.25)\,\Delta^2 f(37)}{2}$$

$$+ \cdots$$

$$= 0.60182 + (0.75)(0.01384) + \frac{(0.75)(-0.25)(-0.00018)}{2}$$

$$+ \frac{(0.75)(-0.25)(-1.25)(-0.00001)}{6}$$

$$= 0.60182 + 0.01038 + 0.000016875 - 0.0000003906$$

$$= 0.6122164844 = 0.61222$$

This compares with the value 0.61221728 in the trigonometric tables.

There was no need to use $\Delta^4 f$; in fact, the fourth difference shows a commonly observed phenomenon: the differences appeared to be approaching zero, then became random looking, and finally started to increase. It is natural for the differences to decrease if the function is reasonably well behaved, as we shall see below. However, the round-off errors begin to grow in their differences. Table 6.4 shows a table of functional values and differences, indicated by the symbol x. One of the functional values has a small error e; however, all the x's represent exact representations. Note how the error e propagates through the difference

Table 6.4

f	Δf	$\Delta^2 f$	$\Delta^3 f$	$\Delta^4 f$	$\Delta^5 f$
x		x		x	
	x		x		x
x		x		x	
	x		x		$x + e$
x		x		$x + e$	
	x		$x + e$		$x - 5e$
x		$x + e$		$x - 4e$	
	$x + e$		$x - 3e$		$x + 10e$
$x + e$		$x - 2e$		$x + 6e$	
	$x - e$		$x + 3e$		$x - 10e$
x		$x + e$		$x - 4e$	
	x		$x - e$		$x + 5e$
x		x		$x + e$	
	x		x		$x - e$
x		x		x	
	x		x		x
x		x		x	

table with increasing amplitude. Problem 6 suggests how this phenomenon of propagation can be used to try to correct a "smooth" table containing an error in one of the entries.

Note also that the difference table "instructs" the user in what order of interpolation is necessary. We simply keep differencing and adding terms until the new terms are so small that they do not affect the round-off. This is in contrast to the Lagrange method, where the degree of polynomial to be used as an interpolation polynomial must be decided a priori.

A third point to be remembered is that one can never exceed the accuracy of the table. It is not possible, by interpolation, to obtain an accurate 6-digit functional value from a 5-digit table. Therefore, results must be rounded back to the table accuracy; in general, the accuracy of the result may be less than that of the table. However, this should not preclude carrying more digits during the computation. That is good computational practice and tends to improve accuracy. The result must not be represented as having more accuracy than warranted, which means that the final result should be rounded properly.

To show that the differences do tend to go to zero, we look at the remainder term in Newton's forward formula:

$$\frac{\alpha(\alpha - 1) \cdots (\alpha - n + 1)(\alpha - n)}{(n + 1)!} f^{(n+1)}(\xi) h^{n+1}$$

If $|\alpha| < 1$ it is easy to show that the numerator of the coefficient is no greater in magnitude than $(n + 1)!$, so that the coefficient is bounded. Now suppose we have a well-behaved function such as $\sin Mx$. For this function the magnitude of the $(n + 1)$st derivative is less than or equal to $|M|^{n+1}$. If $|Mh| < 1$, then the remainder term approaches zero as n increases, which requires that the successive terms in the Newton forward formula decrease with n. The coefficients of the terms do not decrease rapidly, so we are forced to conclude that the differences themselves must be decreasing with the order of the difference.

For a function such as $\sin Mx$, if $M = 1$, the period is 2π. If $|Mh| < 1$, then $|h| < 1/|M|$; in this case, $|h| < 1$, which is less than $\frac{1}{6}$ of the period. Ordinarily, for interpolation in tables, the value of h is likely to be hundreds of times smaller than this, so there are few cases where the differences will not decrease rapidly.

A second interpolation formula is Newton's backward formula (NBF):

$$f(x_n + \alpha h) = f(x_n) + \alpha \nabla f(x_n) + \frac{\alpha(\alpha + 1)}{2!} \nabla^2 f(x_n) + \cdots$$

$$+ \frac{\alpha(\alpha + 1) \cdots (\alpha + n - 1)}{n!} \nabla^n f(x_n) + \frac{\alpha(\alpha + 1) \cdots (\alpha + n)}{(n + 1)!} f^{(n+1)}(\xi) h^{n+1}$$

where the symbols have the same meanings as before and where ξ is somewhere in the interval bounded by the extremes of x_n, x_0, and $x = x_n + \alpha h$ if x is outside the interval (x_0, x_n).

We can use this formula to find $\sin 37°45'$ but not if $37°$ is taken as x_n. Rather, we would be forced to use x_n as $40°$ or greater to obtain very many differences (if, indeed, this is all the table at our disposal). However, this makes α very large in magnitude and tends to degrade the accuracy of the interpolation. NFF and NBF are most suited to interpolation at the beginning or end of a table. For general interpolation in the interior of a table, central differences are superior because they tend to involve data points symmetrically placed with respect to the interpolation point. Central differences, designated by δ, are defined by

$$\delta f(x) = f(x + \frac{h}{2}) - f(x - \frac{h}{2})$$

$$\delta^2 f(x) = f(x + \frac{h}{2}) - 2f(x) + f(x - \frac{h}{2})$$

$$\delta^3 f(x) = f(x + \frac{3h}{2}) - 3f(x + \frac{h}{2}) + 3f(x - \frac{h}{2}) - f(x - \frac{3h}{2})$$

Because the functional values such as $f(x + h/2)$ are generally not available, the commonly used interpolation formulas for central differences involve even-ordered differences or contain terms such as $\delta f(x + h/2)$ and $\delta f(x - h/2)$. The study of these methods takes us too far from our objective, so we refer the reader to any advanced numerical analysis text for further study.

Divided Differences If the base points $\{x_i\}$ are not evenly spaced, we must then resort to using divided differences. These are often useful also in cases where the base points are evenly spaced. The notation is as follows.

First divided difference:

$$f[x_1, x_2] = \frac{f(x_2) - f(x_1)}{x_2 - x_1}$$

Second divided difference:

$$f[x_1, x_2, x_3] = \frac{f[x_2, x_3] - f[x_1, x_2]}{x_3 - x_1}$$

Nth divided difference:

$$f[x_1, \ldots, x_n] = \frac{f[x_2, \ldots, x_n] - f[x_1, \ldots, x_{n-1}]}{x_n - x_1}$$

It is easy to show that $f[x_1, x_2] = f[x_2, x_1]$. Furthermore, it is possible to show that for any higher-ordered divided difference, the symbols (x's) may be permuted in any order without changing the value of the divided difference.

Divided differences lend themselves to the construction of nested interpolating polynomials in an interesting way. Suppose we have a set of points $\{x_i, f(x_i)\}$ to be interpolated by an $(n - 1)$st-degree polynomial. We write the polynomial in the form

$$p(x) = a_0 + a_1(x - x_1) + a_2(x - x_1)(x - x_2) + \cdots$$

$$+ a_{n-1}(x - x_1)(x - x_2) \cdots (x - x_{n-1})$$

Now $p(x_1) = a_0$, which requires that $a_0 = y_1 = f(x_1)$. Then $p(x_2) = a_0 + a_1(x_2 - x_1) = y_1 + a_1(x_2 - x_1) = y_2$, from which we find that

$$a_1 = \frac{y_2 - y_1}{x_2 - x_1} = f[x_1, x_2]$$

Next we have

$$p(x_3) = y_3 = y_1 + f[x_1, x_2](x_3 - x_1) + a_2(x_3 - x_1)(x_3 - x_2)$$

from which we find

$$a_2 = \frac{y_3 - y_1 - f[x_1, x_2](x_3 - x_1)}{(x_3 - x_1)(x_3 - x_2)} = \frac{f[x_1, x_3] - f[x_1, x_2]}{x_3 - x_2}$$

$$= f[x_2, x_1, x_3] = f[x_1, x_2, x_3]$$

Finally,

$$p(x) = f[x_1] + f[x_1, x_2](x - x_1) + f[x_1, x_2, x_3](x - x_1)(x - x_2)$$

$$+ f[x_1, x_2, x_3, x_4](x - x_1)(x - x_2)(x - x_3) + \cdots$$

It is easy to construct this polynomial by first constructing a table of divided differences. We note that $x - x_1$ is a factor in every term but the first, that

$x - x_2$ is a factor in every term but the first two, and so on, which suggests that we should write the polynomial in nested form:

$$p(x) = \{ \cdots [a_{n-1}(x - x_{n-1}) + a_{n-2}](x - x_{n-2}) + \cdots + a_1\}(x - x_1) + a_0$$

This is a useful form because it allows the evaluation of the polynomial to be done with a minimum of computer arithmetic. You are advised to look at the problem concerning the speed of evaluation of polynomials in the exercises (Exercise 7).

POLYCOEFF **and** POLYVAL   The procedure POLYCOEFF and the function POLY-VAL provide the ability to compute the coefficients of an interpolating polynomial of arbitrary degree and to compute values of the polynomial function for an arbitrary abscissa. If these subprograms are used with any main program, POLY-COEFF is called first to compute the polynomial coefficients. This is required only once, provided that the vector of coefficients is not altered or destroyed. POLY-VAL can then be called as often as desired in order to provide one value of the polynomial per call.

It is necessary in the main program to define a data type:

```
TYPE VECTOR = ARRAY[1..NUM] OF REAL;
```

where NUM is an integer to define the expected maximum array size. The call to POLYCOEFF is

```
POLYCOEFF (N, X, Y);
```

where N is the number of points, X is the vector of abscissas of the points, and Y is the vector of ordinates of the points. The original values of Y are destroyed by POLYCOEFF and are replaced by the constants of the $(N - 1)$st-degree polynomial.

POLYVAL is a function, and the call to it involves the name POLYVAL as a variable on the right-hand side of an expression. The parameters are arranged

```
POLYVAL (N, X, Y, V)
```

where N is the number of data points, X is a vector of the N abscissas of the data points, Y is a vector of coefficients produced by POLYCOEFF, and V is a real number representing the abscissa at which the value of the polynomial is to be determined. The polynomial is evaluated in nested form as described in the previous section.

Figure 6.5 contains a graph of a curve known as the *witch of Agnesi*, in this case with the equation (parameter $a = \frac{1}{2}$)

$$y = \frac{1}{1 + x^2}$$

together with the interpolating polynomial of 10th degree that interpolates the points with abscissas $-5, -4, \ldots, 4, 5$, all spaced 1 unit apart. This shows a

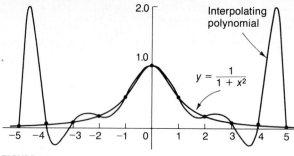

FIGURE 6.5

common problem encountered with interpolating polynomials of high degree when used this way. Polynomials are "polynomialish" and they tend to wiggle up and down. When we want to fit something such as a horizontally asymptotic curve, it is nearly impossible unless we select the abscissas with great care. We can tame the polynomial by choosing the abscissas appropriately, but such a choice may not be at our disposal. The theory is beyond the scope of the book, and furthermore there are other methods of producing better interpolations than are obtained with high-degree polynomials.

Note also that the wiggles in the polynomial interpolations shown in Figure 6.1 and Figure 6.5 occur near the ends of the interval over which the polynomial is used. This is characteristic behavior of polynomials, and it shows why the central-difference interpolating formulas are used for interpolation in tables whenever possible. The polynomial simply fits better in the middle.

In Chapter 9 we will look at other methods of representing functions by polynomials in which we try to control the maximum differences between the values of the function and the polynomial. As we will see there, this requires that we give up the fixed points at which $p(x) = f(x)$ in favor of bounds on the maximum error $|f(x) - p(x)|$. The effect this has can be seen in Figure 9.2, which shows a *minimax approximation* that corresponds to the interpolation polynomial shown in Figure 6.5.

Inverse Interpolation   Occasionally it is necessary to find the value of the abscissa for a point when the functional value is given and the functional value is not one in the table. When logarithms were commonly used for multiplication, division, and taking roots, this was known as finding an antilog. The problem is really no different from that of any other interpolation. The values of the function now become the abscissas and the x's become the ordinates. Divided differences are generally necessary because the functional values in the table are almost never uniformly spaced.

Newton's divided-difference polynomial can be used for this purpose. Example 6.2 shows the use of inverse interpolation to find the angle whose sine is 0.63944.

Example 6.2   Find the angle whose sine is 0.63944.

The angle is between 39° and 40°. We construct the divided-difference table, Table 6.5. The first entry in the column for $f[x, x]$ is $(37 - 36)/(0.60182 - 0.58779)$ = 71.27584, for instance.

Then

$$f(y) = f[y_1] + f[y_1, y_2](y - y_1) + f[y_1, y_2, y_3](y - y_1)(y - y_2) + \cdots$$

with $y_1 = 0.62932$, $y_2 = 0.64279$, ..., and $f[y_1] = 39$, $f[y_1, y_2] = 74.23905$, ..., and $y = 0.63944$, or

$$f(y) = 39 + 74.23905(0.63944 - 0.62932)$$
$$+ 41.84368(0.63944 - 0.62932)(0.63944 - 0.64279)$$
$$+ 48.61492(0.63944 - 0.62932)(0.63944 - 0.64279)(0.63944 - 0.65606)$$
$$+ \cdots$$
$$= 39 + 0.751299186 - 0.0014185844 + 0.0000273921 + \cdots$$
$$= 39.74990799 = 39.750$$

This is compatible with the accuracy of the table (5 significant digits). Note the noisy variation in the third divided differences; this is relatively no worse than the noise on the third differences of Table 6.2.   □

Table 6.5 Table to Illustrate Inverse Interpolation on the Sine Function

Sine X	X (deg)	F[x, x]	F[x, x, x]	F[x, x, x, x]
0.58779	36			
		71.27584		
0.60182	37		35.10944	
		72.25434		-11.74139
0.61566	38		34.62182	
		73.20644		83.95753
0.62932	39		38.06156	
		74.23905		93.61683
0.64279	40		41.84368	
		75.35795		48.61492
0.65606	41		43.77904	
		76.51109		52.46417
0.66913	42		45.83616	
		77.70008		
0.68200	43			

6.4 Splines

Rather than trying to use a high-degree polynomial for interpolation, we can try using pieces of lower-degree polynomials, joined end to end at the interpolation points. We call these points *knots*. We can impose conditions of continuity and smoothness on the composite function, depending upon the degree of polynomial we wish to use between the knots. For example, a set of straight-line segments (first-degree polynomials) allows us to impose a condition of continuity, which means that we force the ends of the line segments to meet at the knots, as shown in Figure 6.3. Thus we have a continuous function but one that is not very smooth; its first derivative is discontinuous at the knots because the slopes of the straight-line segments that join at a knot are different.

The use of pieces of parabola (second-degree polynomials) allows us to impose continuity conditions and slope conditions at the knots. This causes the composite function to have a much smoother appearance, but the requirement that the slopes match at the knots sometimes forces the parabolas into forming wild oscillations, as shown in Figure 6.6. Perhaps the most commonly used spline is the cubic spline, where we can impose conditions of continuity, slope, and second derivative, which amounts to curvature. Cubics can have inflection points, which relieves the oscillation problem found in quadratic splines. This is obvious if we compare Figure 6.7 with Figure 6.6.

Matching the second derivatives at the knots provides another smoothness effect: If the curve were a highway, a second-degree spline curve would appear smooth, but it could have instantaneous changes in curvature. These changes would require instantaneous, large changes in the angle of the steering wheel to keep the car on the road. This would not be the case with a cubic spline. Highway engineers recognize this fact and no longer connect a north-south segment of road to an east-west segment with a quarter-circle; a transition curve is

FIGURE 6.6   Oscillations exhibited by a quadratic spline.

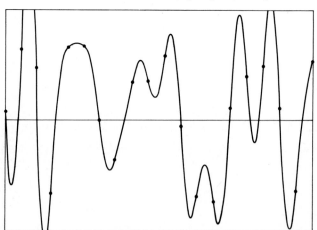

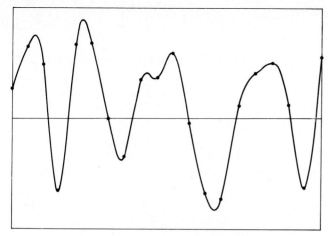

FIGURE 6.7 Cubic spline interpolation.

used to go from the straight piece to the circular piece to allow the driver to have better steering control. We could use even higher-degree splines and match first, second, and third (or higher) derivatives at the knots, but for most applications this is unnecessary. Cubic splines are adequate.

Suppose we have $n + 1$ knots $\{(x_i, y_i)\}$, which we would like to interpolate with n cubic curve segments having, in general, $4n$ constants to be determined. Each segment must pass through two points, which gives us $2n$ equations to be used in solving for the $4n$ constants. There are $n - 1$ internal knots at which the segments join, and if we impose conditions on both the first and second derivatives, we have $2(n - 1)$ additional equations. Thus we have a total of $4n - 2$ equations to be used in solving for $4n$ constants, which means that we can impose two arbitrary conditions to produce two more equations. By choosing the way in which we write the equations, we do not need $4n$ equations. This is shown in the next section.

6.4.1 Natural Splines

Let the spline $S(x)$ be represented by a set of cubic spline segments $s_i(x)$, $x_i \leqslant x \leqslant x_{i+1}$, where the end knots are at x_0 and x_n. We impose the conditions

$$s_i(x_i) = y_i \qquad \text{for } i = 0, 1, \ldots, n - 1$$

$$s_i(x_{i+1}) = y_{i+1} \qquad \text{for } i = 0, 1, \ldots, n - 1$$

$$s'_{i-1}(x_i) = s'_i(x_i) \qquad \text{for } i = 1, \ldots, n - 1$$

$$s''_{i-1}(x_i) = s''_i(x_i) \qquad \text{for } i = 1, \ldots, n - 1$$

Note that we have a total of $4n - 2$ equations. The natural spline is defined by setting the second derivatives to zero at the two endpoints:

$$s''_0(x_0) = 0$$

$$s''_{n-1}(x_n) = 0$$

which means that there is no curvature at the extreme ends of the spline.

If we use a thin strip of metal, plastic, or wood to construct a physical spline, we can use it just as a draftsperson does to fit a set of points for drawing a curve. If there are no torques applied to the ends (think of the spline as being slotted so that pins representing the knots cannot exert either tension forces or torques), the ends beyond the first and last knots will be straight. This is a natural spline. It can be shown that the linearized differential equation for a simple beam subject to spline-type constraints has solutions that are third-degree in the independent variable x.

To find the constants for the spline segments, we use a special notation that simplifies the final solution. First, we define $S(x)$ as a function that is represented by the spline segments:

$$S(x) = s_i(x) \qquad \text{if } x_i \leqslant x \leqslant x_{i+1}$$

Then we set $S''(x_i) = z_i$, where the z_i will form a vector of coefficients that will need to be determined (these are the second derivatives at the knots). By using the special form, we are able to generate the spline segments from the vector of x_i (the set of abscissas of the knots), the vector of y_i (the set of ordinates of the knots), and the vector of z_i.

Let $h_i = x_{i+1} - x_i$, and choose $s''_i(x)$ in the form

$$s''_i(x) = \frac{(x_{i+1} - x)z_i + (x - x_i)z_{i+1}}{h_i}, \qquad i = 0, 1, 2, \cdots, n-1$$

Note that $s''_i(x_i) = z_i$ and $s''_i(x_{i+1}) = z_{i+1}$, as required. We now find the antiderivative twice using a special form

$$s_i(x) = \frac{(x_{i+1} - x)^3 z_i + (x - x_i)^3 z_{i+1}}{6h_i} + b_i(x_{i+1} - x) + a_i(x - x_i)$$

where the a_i and b_i are constants of integration. Now $s_i(x_i) = y_i$ and $s_i(x_{i+1}) = y_{i+1}$, so that

$$y_i = \frac{h_i^2 z_i}{6} + b_i h_i$$

$$y_{i+1} = \frac{h_i^2 z_{i+1}}{6} + a_i h_i$$

Thus we can write (show this as a problem)

$$s_i(x) = \frac{(x_{i+1} - x)^3 z_i + (x - x_i)^3 z_{i+1}}{6h_i} + \frac{(x_{i+1} - x)y_i + (x - x_i)y_{i+1}}{h_i}$$

$$+ \frac{h_i}{6}[(x_{i+1} - x)z_i + (x - x_i)z_{i+1}]$$

We now write $s'_i(x_{i+1}) = s'_{i+1}(x_{i+1})$, which allows us to find

$$h_i z_i + 2(h_{i+1} + h_i)z_{i+1} + h_{i+1}z_{i+1}$$
$$= 6\left[\frac{y_{i+2} - y_{i+1}}{h_{i+1}} - \frac{y_{i+1} - y_i}{h_i}\right], \qquad i = 0, 1, 2, \cdots, n-2$$

This is a tridiagonal system, which we can solve for the z_i. The only problem is that there are $n - 1$ equations and $n + 1$ z's. For the natural spline we set $z_0 = 0$ and $z_n = 0$, which leaves $n - 1$ z's to be solved for in a tridiagonal system.

One property of the natural spline $S(x)$ is that over a given set $\{(x_i, y_i)\}$ this spline minimizes the integral

$$\int_{x_0}^{x_n} (S''(x))^2\, dx$$

A proof is beyond the scope of this book. Because we realize that the second derivative is so closely related to curvature, we see that minimization of the integral tends to minimize the oscillatory behavior of the spline. It will not exhibit the wild behavior of a high-degree polynomial through the same set of points. Figure 6.8 shows the same witch curve and the same set of interpolation points that were used in preparing Figure 6.5. The fit is remarkably better (although not perfect).

Because cubic splines are so well behaved, they are often used in computer graphics routines for curve plotting. The curve is often defined by a discrete set of points, and in that case, the spline routines can be used to generate a smooth set of "fill-in" points in order to generate a smooth curve. The spline never springs any surprises.

FIGURE 6.8   Witch curve interpolated by cubic spline.

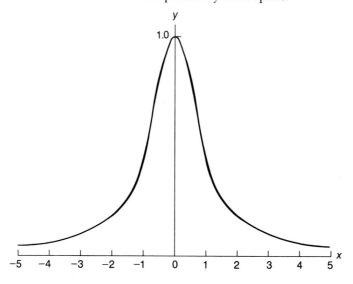

6.4.2 Other Splines

The natural spline condition may be unnatural for some problems, and it is possible to use other conditions to supply the two missing equations necessary to solve for all $4n$ constants. One possibility, which might be a physical constraint in a problem, is to put slope conditions on the endpoints (or on interior points, for that matter). A common constraint might be called *preservation of local curvature*. This requires that we estimate the second derivative from the first three and last three points, using divided differences. These estimated second derivatives are then imposed upon z_0 and z_n in order to reduce the number of independent constants to $n - 1$. Modification of the system equations to be solved is left as a project problem, along with modification of the programs for the solution.

6.4.3 Procedure CUBIC_SPLINE and Function SPLEVAL

CUBIC_SPLINE is a PASCAL procedure that determines the necessary spline coefficients to be used in function SPLEVAL to do a natural cubic spline interpolation through a set of points (knots). The details of the derivation were discussed earlier.

For these routines to work properly, the main program must contain a data-type declaration

```
TYPE  VECTOR = ARRAY[1..NUM] OF REAL;
```

where NUM is an integer number for the maximum expected number of knots. The call to CUBIC_SPLINE is

```
CUBIC_SPLINE (N, B, C, D, Y, Z);
```

where

N = actual number of knots ($N \le$ NUM)

B = vector of abscissas of the knots in increasing order

Y = vector of ordinates of the knots in the same order as B

C, D = vectors of constants generated by CUBIC_SPLINE

(C and D are local vector variables not used by SPLEVAL)

Z = vector of constants passed to SPLEVAL along with B and Y

Procedure CUBIC_SPLINE calls procedure TRIDIAG (Chapter 5) to solve for the vector Z; the procedure is incorporated within CUBIC_SPLINE.

Function SPLEVAL is called in the usual function manner, with the parameters arranged as shown:

```
FUNCTION SPLEVAL (N, B, Y, Z, V)
```

where N, B, Y, and Z are as described for CUBIC_SPLINE. V is the abscissa of the point at which the spline is to be evaluated. The value is returned associated with the variable name SPLEVAL.

Upon entry to SPLEVAL, X is tested to see which spline segment is to be used. This is done by a linear search, which is not greatly efficient. If X > B [N], the last spline segment is used for an extrapolation. If X < B [0], then the first segment is used for extrapolation. The evaluation of the polynomial is in nested form.

If SPLEVAL is to be used extensively for graphing, it is advisable to modify SPLEVAL so that an index is kept on the segment number of the most recent access. If X is not in that segment on the next access, a binary search can be instituted in order to find the new segment more quickly than with a linear search. This modification is left as a project problem.

6.5 Least Squares

We now look at the problem of interpolating a set of points where the points are noisy, or contain measurement errors. Frequently the errors tend to be mainly in one coordinate. The measurement of the other coordinate may be much more accurate. For instance, time can be measured with great accuracy, but other kinds of quantities may have their measurements disturbed by electrical noise or be inherently inaccurate.

Often in experimental situations there may be underlying theory that suggests the kind of function to be used in fitting the data. For example, the radioactive decay of a single species of atom is known to obey a physical law that requires the use of an exponential function of the form

$$N = N_0 e^{-at}$$

to represent the number N of atoms present at any time t. The problem is to determine an appropriate value for N_0 and for a from a set of measurements. In other cases there may be no theory on which to rely in selecting a function to represent the data. In such circumstances a polynomial is often used for want of anything better; sometimes a rational function is used if there seem to be horizontal or vertical asymptotes.

6.5.1 Linear Regression

The simplest case of curve fitting is that of a linear polynomial. If the function is $y = a + bx$ and we assume that the x's are exact and that the y's contain all the error, then we call the method of least squares a *linear regression of y on x*. If we assume that the y's are exact and that the x's contain the errors, then we can produce a *linear regression of x on y*. Since the details of both are the same, we deal only with the former case.

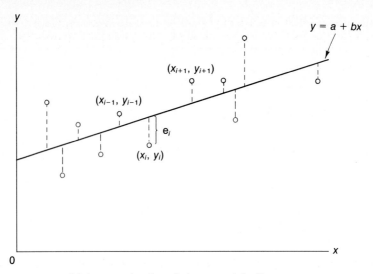

FIGURE 6.9   Noisy set of points fit by a straight line.

Figure 6.9 shows a set of points and a straight line through the set of points. We try to choose the line in such a way as to minimize some measure of the deviations of the data points and the points on the line having the same abscissas. We use a sum so that every point is able to exert its influence, and we use squares of the deviations in order to make the mathematics relatively easy. Squares make the terms in the sum all positive, so that we can try to find a minimum using the calculus method of setting derivatives equal to zero. (Higher, even powers simply complicate matters, and absolute values have discontinuous derivatives at the minimum points.)

Suppose we have a set of m data points $\{(x_i, y_i)\}$, where m is greater than 2, and a set of weights $\{w_i\}$ corresponding to the points. The weights express our confidence in the accuracy of the points; if we think that they are equally accurate, then we can set them all to 1. However, there can be cases where part of the data are collected with different equipment or where a measuring device is less accurate for large (or small) values of the quantity being measured.

We wish to find the "best" values of a and b for the equation $y = a + bx$, where this is assumed to model the data. The deviation at each point is

$$e_i = a + bx_i - y_i$$

We now form the weighted sum of squares of the deviations

$$F(a, b) = \sum_{i=1}^{m} w_i(e_i)^2$$

or

$$F(a, b) = \sum_{i=1}^{m} w_i(a + bx_i - y_i)^2$$

where we note that $F(a, b)$ is a function of the yet-to-be-determined a and b. We now find the partial derivatives of F with respect to a and b and set these to zero.

$$\frac{\partial F}{\partial a} = 2\sum_{i=1}^{m} w_i(a + bx_i - y_i) = 0$$

$$\frac{\partial F}{\partial b} = 2\sum_{i=1}^{m} w_i(a + bx_i - y_i)x_i = 0$$

These equations can be turned into normal form

$$a\sum_{i=1}^{m} w_i + b\sum_{i=1}^{m} x_iw_i = \sum_{i=1}^{m} y_iw_i$$

$$a\sum_{i=1}^{m} x_iw_i + b\sum_{i=1}^{m} x_i^2w_i = \sum_{i=1}^{m} x_iy_iw_i$$

If the points are equally weighted, the w's can all be set to 1. In this case we have

$$ma + b\sum_{i=1}^{m} x_i = \sum_{i=1}^{m} y_i$$

$$a\sum_{i=1}^{m} x_i + b\sum_{i=1}^{m} x_i^2 = \sum_{i=1}^{m} x_iy_i$$

Example 6.3

We have a set of points, $(1, 1)$, $(2, 1.5)$, $(3, 1.75)$, and $(4, 2)$, for which we would like to find the regression line of y on x, or the best-fitting straight line in the sense of least squares.

$$m = 4$$

$$\sum_{i=1}^{4} x_i = 10$$

$$\sum_{i=1}^{4} (x_i)^2 = 30$$

$$\sum_{i=1}^{4} y_i = 6.25$$

$$\sum_{i=1}^{4} x_iy_i = 17.25$$

We must solve

$$\begin{bmatrix} 4 & 10 \\ 10 & 30 \end{bmatrix} \begin{bmatrix} a \\ b \end{bmatrix} = \begin{bmatrix} 6.25 \\ 17.25 \end{bmatrix}$$

to find $a = 0.75$ and $b = 0.325$. The regression equation is

$$y = 0.75 + 0.325x$$

See Figure 6.10.   □

Note that $F(a, b)$ is minimum for this particular set of values of a and b for the specific set of data points, as the reader can check with a calculator.

_____

Example 6.4

Suppose we would like to give five times as much emphasis to the middle two points as to the two endpoints. Then we set the weights as $w_1 = w_4 = 1$ and $w_2 = w_3 = 5$.

$$\sum_{i=1}^{4} w_i = 12$$

$$\sum_{i=1}^{4} w_i x_i = 30$$

$$\sum_{i=1}^{4} w_i (x_i)^2 = 82$$

$$\sum_{i=1}^{4} w_i y_i = 19.25$$

$$\sum_{i=1}^{4} w_i x_i y_i = 50.25$$

Again we must solve the normal equations

$$\begin{bmatrix} 12 & 30 \\ 30 & 82 \end{bmatrix} \begin{bmatrix} a \\ b \end{bmatrix} = \begin{bmatrix} 19.25 \\ 50.25 \end{bmatrix}$$

to find $a = 0.8452$ and $b = 0.3036$. Note in Figure 6.11 that the line now passes very close to the middle two points, as expected.   □

A measure of the goodness of fit is the value of F/m, where $F = F(a, b)$, which in the uniformly weighted case is the mean square error. In the first example the sum of the squares of the errors, or deviations, is 0.01875, which makes the mean square error equal to 4.6875×10^{-3}. Statisticians have a related measure called the *standard error of estimate*, which is $(F/(m - 2))^{1/2}$. For this example the standard error of estimate is 9.682×10^{-2}.

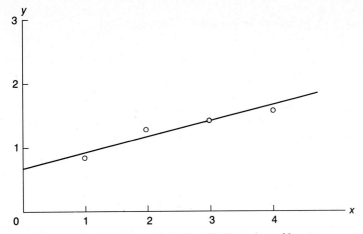

FIGURE 6.10 Best-fitting straight line in the sense of least squares (equally weighted data).

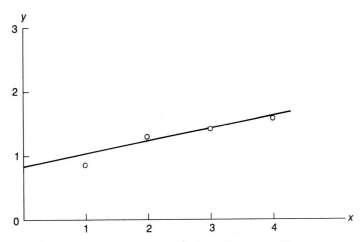

FIGURE 6.11 Best-fitting straight line in the sense of least squares (unequally weighted data).

6.5.2 Multiple Regression

Sometimes several variables may affect the value of another variable in a linear way, so that we can write

$$y = a + b_1x_1 + b_2x_2 + \cdots + b_nx_n$$

as a model equation for that relationship. To find the coefficients for equally weighted points, we again form the sum of squares of the deviations

$$F(a, b_1, \ldots, b_n) = \sum_{i=1}^{m} [a + b_1 x_{1i} + b_2 x_{2i} + \cdots + b_n x_{ni} - y_i]^2$$

Next we differentiate F with respect to each of the coefficients and equate these partial derivatives to zero to find the normal equations

$$ma + b_1 \sum_{i=1}^{m} x_{1i} + b_2 \sum_{i=1}^{m} x_{2i} + \cdots + b_n \sum_{i=1}^{m} x_{ni} = \sum_{i=1}^{m} y_i$$

$$a \sum_{i}^{m} x_{1i} + b_1 \sum_{i} x^2_{1i} + b_2 \sum_{i} x_{2i} x_{1i} + \cdots + b_n \sum_{i} x_{ni} x_{1i} = \sum_{i} y_i x_{1i}$$

$$a \sum_{i=1}^{m} x_{ni} + b_1 \sum_{i} x_{1i} x_{ni} + b_2 \sum_{i} x_{2i} x_{ni} + \cdots + b_n \sum_{i} x^2_{ni} = \sum_{i=1}^{m} y_i x_{ni}$$

or in matrix form

$$A\mathbf{b} = \mathbf{y}$$

where

$$A = \begin{bmatrix} m & \Sigma x_{1i} & \Sigma x_{2i} & \cdots & \Sigma x_{ni} \\ \Sigma x_{1i} & \Sigma x^2_{1i} & \Sigma x_{1i} x_{2i} & \cdots & \Sigma x_{1i} x_{ni} \\ \Sigma x_{2i} & \Sigma x_{2i} x_{1i} & \Sigma x^2_{2i} & \cdots & \Sigma x_{2i} x_{ni} \\ \cdot & \cdot & \cdot & & \cdot \\ \cdot & \cdot & \cdot & & \cdot \\ \cdot & \cdot & \cdot & & \cdot \\ \Sigma x_{ni} & \Sigma x_{ni} x_{1i} & \Sigma x_{ni} x_{2i} & \cdots & \Sigma x^2_{ni} \end{bmatrix}$$

$$\mathbf{b} = \begin{bmatrix} a \\ b_1 \\ b_2 \\ \cdot \\ \cdot \\ \cdot \\ b_n \end{bmatrix}$$

and

$$\mathbf{y} = \begin{bmatrix} \Sigma y_i \\ \Sigma x_{1i} y_i \\ \Sigma x_{2i} y_i \\ \cdot \\ \cdot \\ \cdot \\ \Sigma x_{ni} y_i \end{bmatrix}$$

Note that this is an $n + 1 \times n + 1$ linear system, which can be solved by the methods of Chapter 5.

If $n = 2$, the equation $y = a + b_1x_1 + b_2x_2$ has the obvious interpretation of being a plane surface above the x_1x_2 plane. If $n > 2$, the meaning is clear but the visualization is not so clear. By analogy with the $n = 2$ case we say that the equation represents a plane surface in a multidimensional space.

6.5.3   Nonlinear Regression

We can also fit polynomials of higher degree than first to sets of data. Let $y = a + bx^2 + cx^3 + \cdots + qx^r$, and let there be m points (x_i, y_i) and weights w_i. Form the sum of squares of the deviations, as before

$$F(a, b, \cdots, q) = \sum_{i=1}^{m} (a + bx_i + cx_i^2 + \cdots + qx_i^r - y_i)^2 w_i$$

Set the partial derivatives of F with respect to the coefficients equal to zero to find the normal equations

$$a\sum_i w_i + b\sum_i w_ix_i + c\sum_i w_ix_i^2 + \cdots + q\sum_i w_ix_i^r = \sum_i w_iy_i$$

$$a\sum_i w_ix_i + b\sum_i w_ix_i^2 + c\sum_i w_ix_i^3 + \cdots + q\sum_i w_ix_i^{r+1} = \sum_i w_ix_iy_i$$

$$\cdot$$
$$\cdot$$
$$\cdot$$

$$a\sum_i w_ix_i^r + b\sum_i w_ix_i^{r+1} + c\sum_i w_ix_i^{r+2} + \cdots + q\sum_i w_ix_i^{2r} = \sum_i w_ix_i^ry_i$$

or, in matrix form,

$$\begin{bmatrix} \Sigma w_i & \Sigma w_ix_i & \cdots & \Sigma w_ix_i^r \\ \Sigma w_ix_i & \Sigma w_ix_i^2 & \cdots & \Sigma w_ix_i^{r+1} \\ \cdot & & & \\ \cdot & & & \\ \cdot & & & \\ \Sigma w_ix_i^r & & \cdots & \Sigma w_ix_i^{2r} \end{bmatrix} \begin{bmatrix} a \\ b \\ \cdot \\ \cdot \\ \cdot \\ q \end{bmatrix} = \begin{bmatrix} \Sigma w_iy_i \\ \Sigma w_ix_iy_i \\ \cdot \\ \cdot \\ \cdot \\ \Sigma w_ix_i^ry_i \end{bmatrix}$$

Unfortunately, for values of r in the range of six or more, the condition of the matrix is often very poor. For these cases the use of orthogonal polynomials is much superior, as will be discussed later.

How high should be the degree of the polynomial used? In general, it should be as low as possible. One method of determination of the degree is to compute the mean square error or the standard error of estimate for degrees 1, 2, 3, . . . , and observe how the error seems to behave. Going to a degree higher than necessary will generally not improve the error very much, if at all.

In this case the standard error of estimate is $\sqrt{F/(m - r - 1)}$, where m is the number of points, r is the degree of the polynomial, and F is the sum of squared deviations.

Example 6.5

Table 6.6 shows a set of data points manufactured by making each y equal to x^3 plus a random number uniformly distributed between 0 and 1. Table 6.7 shows the results of the regressions for polynomials of degrees 1 through 4.

Note that there were reductions in the standard error of estimate by substantial amounts in going from one model to another until the correct model was reached. In going on to the fourth-degree equation, the error did not improve (the fact that it increased slightly is not significant). We should also observe that the coefficient of the fourth-power term in the fourth-degree equation is very small, which should tell us something.   □

Table 6.6   Random Data Points

x_i	y_i
1.0	1.715
1.5	4.058
2.0	8.482
2.5	16.624
3.0	27.647
3.5	43.007
4.0	64.369
4.5	91.712
5.0	125.135

Table 6.7   Results of Regression with Polynomials of Degrees 1 through 4

Degree	Polynomial Coefficients	Standard Error of Estimate
1	-46.9522	14.99
	29.8266	
2	18.8482	1.9671
	-24.0101	
	8.9728	
3	0.47531	0.27997
	0.33528	
	-0.13775	
	1.01228	
4	-0.01892	0.31196
	1.23157	
	-0.67422	
	1.14131	
	-0.01075	

It is necessary to have more points than coefficients. It is left as an exercise to show what happens when the number of points is the same as or fewer than the number of coefficients.

6.5.4 Orthogonal Polynomials

As we have seen, the normal equations for fitting high-degree least-squares polynomial curves tend to be ill-conditioned. We can see how bad the normal equations are by calculating the last two rows in the matrix for the case of a fifth-degree least-squares polynomial fit using $x_1 = 1970$, $x_2 = 1971$, ..., $x_{11} = 1980$. The ratios of the elements in the last two rows of the normal matrix are

$$\frac{a_{6,1}}{a_{5,1}} = 1975.02025$$

$$\frac{a_{6,2}}{a_{5,2}} = 1975.02532$$

$$\frac{a_{6,3}}{a_{5,3}} = 1975.03038$$

$$\frac{a_{6,4}}{a_{5,4}} = 1975.03544$$

$$\frac{a_{6,5}}{a_{5,5}} = 1975.04050$$

$$\frac{a_{6,6}}{a_{5,6}} = 1975.04557$$

This shows that the sixth row is so close to being a multiple of the fifth row that we should not be surprised when we find that the condition number of the normal matrix is greater than 10^{20}. With such a condition number we should have grave doubts about the accuracy of any solution that we might compute. The root of the problem is that the functions 1, x, x^2, x^3, ..., are far from being an orthogonal set of functions.

To alleviate that condition let us introduce, instead of powers of x, a set of polynomials $P_k(x)$, where k is the degree of the polynomial, having the property that

$$\sum_{i=1}^{m} P_j(x_i)P_k(x_i) = 0 \qquad j \neq k$$

where m is the number of x's. Unfortunately, these polynomials are specific to the set of x_i, but this is the price we have to pay for improved accuracy. For the moment we defer discussing the procedure for constructing the P's.

We now express our desired function $y(x)$ as

$$y(x) = a_0 P_0(x) + a_1 P_1(x) + \cdots + a_n P_n(x)$$

and determine the a's by least squares as before. We will assume that $m > n + 1$; if $m = n + 1$, then the problem becomes one of finding an interpolating polynomial. We write

$$F(a_0, a_1, \ldots) = \sum_{i=1}^{m} [a_0 P_0(x_i) + \cdots + a P_n(x_i) - y_i]^2$$

and try to find the minimum of F with respect to the a's by setting the partial derivatives equal to zero:

$$\frac{\partial F}{\partial a_j} = 2\sum_{i=1}^{m} [a_0 P_0(x_i) + \cdots + a_n P_n(x_i) - y_i] P_j(x_i) = 0$$

When we break up the expression into separate sums, we find that all terms of the form

$$\sum_{i=1}^{m} a_k P_k(x_i) P_j(x_i)$$

are zero except for the one in which $k = j$. Thus we find that

$$a_j \sum_{i=1}^{m} P_j^2(x_i) = \sum_{i=1}^{m} y_i P_j(x_i)$$

From this we see that we do not have to solve a linear system at all in order to find the value of a_j. (It is also correct to say that the normal equation system is diagonal.)

Forsythe [1957] has shown that the polynomials are most easily constructed recursively. We define $P_{-1}(x) = 0$, $P_0(x) = 1$, and

$$P_{k+1}(x) = (x - \alpha_k)P_k(x) - \beta_k P_{k-1}(x), \qquad k > 0$$

If we let

$$\sum_{i=1}^{m} P_k^2(x_i) = \gamma_k$$

then we can show (left as an exercise) that

$$\alpha_0 = \left(\frac{1}{m}\right) \sum_{i=1}^{m} x_i$$

$$\alpha_k = \left(\frac{1}{\gamma_k}\right) \sum_{i=1}^{m} x_i P^2(x_i)$$

$$\beta_k = \frac{\gamma_k}{\gamma_{k-1}}$$

Example 6.6

Repeat the regression problem of Example 6.5, in which the points were (1, 1), (2, 1.5), (3, 1.75), and (4, 2), using orthogonal polynomials.

$$\alpha_0 = \left(\frac{1}{m}\right) \sum_{i=1}^{m} x_i = \left(\frac{1}{4}\right)(10) = 2.5$$

$P_0(x) = 1.0$ (by definition)

$$\gamma_0 = \sum_{i=1}^{4} P_0^2(x_i) = 4 = m$$

$$P_1(x) = (x - \alpha_0)P_0(x) = (x - 2.5)1 = x - 2.5$$

$$\gamma_1 = \sum_{i=1}^{4} P_1^2(x_i)$$

$$= (1 - 2.5)^2 + (2 - 2.5)^2 + (3 - 2.5)^2 + (4 - 2.5)^2 = 5.0$$

$$\beta_1 = \frac{\gamma_1}{\gamma_0} = \frac{5}{1} = 5$$

$$\alpha_1 = \left(\frac{1}{5}\right) \sum_{i=1}^{4} x_i P^2(x_i)$$

$$= \left(\frac{1}{5}\right)[1(1 - 2.5)^2 + 2(2 - 2.5)^2 + 3(3 - 2.5)^2 + 4(4 - 2.5)^2] = 2.5$$

$$P_2(x) = (x - 2.5)P_1(x) - 5P_0(x)$$

$$= (x - 2.5)^2 - 5 = x^2 - 5x + 1.25$$

We now find the a coefficients.

$$a_0 = \left(\frac{1}{\gamma_0}\right) \sum_{i=1}^{4} y_i P_0(x_i) = \left(\frac{1 + 1.5 + 1.75 + 2}{4}\right)$$

$$= 1.5625$$

$$a_1 = \left(\frac{1}{\gamma_1}\right) \sum_{i=1}^{4} y_i P(x_i)$$

$$= \left(\frac{1}{5}\right)[1(1 - 2.5) + 1.5(2 - 2.5) + 1.75(3 - 2.5) + 2(4 - 2.5)]$$

$$= \frac{1.625}{5} = 0.325$$

The linear regression line is

$$y(x) = a_0 P_0(x) + a_1 P_1(x)$$

$$= (1.5625)(1) + (0.325)(x - 2.5) = 0.75 + 0.325x$$

which is the same as before.   □

6.5.5   Orthogonal Polynomial Routines

The program listings in this chapter include four functions, or procedures, that generate polynomials orthogonal over a set of points and determine the coefficients for a least-squares curve fit. For all these, the calling program must contain three type declarations:

```
TYPE COEFVECTOR = ARRAY[0..N] OF REAL;
```

where N is the maximum degree of orthogonal polynomial desired;

```
TYPE DATAVECTOR = ARRAY[1..M] OF REAL;
```

where M is the maximum number of data points expected; and

```
TYPE MATRIX = ARRAY[0..N,1..M] OF REAL;
```

The matrix is used in two different routines and is used to hold the values of all of the $P_k(x_i)$ for all possible k and i.

Procedure ORTHOG produces the set of orthogonal polynomials. The call to ORTHOG is

```
ORTHOG (M, N, X, ALPHA, BETA, GAMMA, PX);
```

where M is the number of data points, N is the highest-degree polynomial desired, X is a DATAVECTOR containing the abscissas of the data points, ALPHA, BETA, and GAMMA are COEFVECTORs, which return the values of those coefficients, and PX is the MATRIX described earlier: the value of every P at every x.

Procedure FIND_COEFFS finds the set of polynomial coefficients (the a's in the previous development) from a set of y's whose values are defined for the x's. The call to this procedure is

```
FIND_COEFFS (M, N, PX, GAMMA, Y, A);
```

where M is the number of data points, N is the degree of the highest-degree polynomial used, PX is the MATRIX of values found by ORTHOG, GAMMA is the COEFVECTOR of values of the γ's (also found by ORTHOG), Y is the DATAVECTOR of y's corresponding to the x's, and A is the return COEFVECTOR containing the set of a's. The call to FIND_COEFFS may use an N smaller than the one used in the call to ORTHOG, but it cannot be larger.

Procedure STD_ERROR produces the standard error of estimate for the polynomial fit. The call is

```
STD_ERROR (M, N, PX, Y, A, STERROR);
```

where M and N are the number of data items and the degree of the highest-degree polynomial, PX is the MATRIX of values of the polynomials at all of the x's, as found in ORTHOG, Y is the DATAVECTOR of y values, A is the COEFVECTOR of a coefficients, as found by FIND_COEFFS, and STERROR is a COEFVECTOR with the standard error of estimate for each degree of polynomial. For a zero-degree polynomial, $y(x) = a_0 P_0(x) = a_0$, the standard error of estimate is found in STER–ROR [0]; for the third-degree polynomial approximation

$$y(x) = a_0 P_0(x) + a_1 P_1(x) + a_2 P_2(x) + a_3 P_3(x)$$

the standard error of estimate is found in STERROR [3].

Function P evaluates any of the individual polynomials recursively. The call, which is a standard kind of function call, is

P (K, ALPHA, BETA, V)

where K is the degree of the particular polynomial desired, ALPHA and BETA are the COEFVECTORs for α and β produced by procedure ORTHOG, and V is the abscissa at which the polynomial is to be evaluated. The value of the polynomial is returned associated with the name of the function.

The use of the function P together with the set of coefficients (the A's, found by FIND_COEFFS) is not the most efficient way to evaluate a least-squares polynomial of the form

$$A_0 P_0(x) + A_1 P_1(x) + \cdots + A_k P_k$$

because the recursive evaluation of each of the higher-degree polynomials requires, in each case, the evaluation of all the lower-degree polynomials. It is left as a project problem (Project Problem 6.8) to revise function P and incorporate it in a new function that is more efficient in evaluating the sum of the orthogonal polynomials.

IMSL **Routines** RLFOTH **and** RLFOTW   These closely related FORTRAN routines are orthogonal polynomial least-squares data-fitting procedures. RLFOTW differs from RLFOTH in that it allows for a vector of weights to be associated with the data points. The routines use the method developed by Forsythe [1957] discussed earlier in the chapter.

The call to RLFOTH is

CALL RLFOTH (X, Y, N, RSQ, MD, ID, P, C, S, A, B, IER)

where X and Y are the real arrays containing the abscissas and ordinates of the N data points. MD is an input integer variable specifying the maximum allowable degree of polynomial that the user wishes. ID is a return variable specifying the highest degree of polynomial actually used (RSQ imposes another limit, as we will see later).

Both RLFOTH and RLFOTW rescale the x variable to the range $[-2, 2]$. This affects all coefficients except for the constant term. Upon return from these routines, the real array C contains the coefficients of the orthogonal polynomials (not the coefficients of the powers of x inside the polynomials). This array must be dimensioned at least MD + 3, which allows room for the MD + 1 coefficients plus the two scaling constants for x. If x' is the scaled value (between -2 and $+2$) of x, then

$$x' = ax + b$$

where a is stored in C (ID + 2) and b is stored in C (ID + 3).

The real arrays A and B must have dimension of at least MD. Upon return, these arrays contain the α's and the β's of the polynomials. The array P is a real work array with dimension at least 2N.

The real variable RSQ is set by the user in the range (0, 100). The number is 100 times the minimum coefficient of determination that the user is willing to accept, where the coefficient of determination is the square of the correlation coefficient. The user may set a value (e.g., 90) and the routine will successively add new polynomial terms to the fit until the coefficient of determination exceeds 0.90, until MD is reached, or until there are not enough points for the degree desired. If the user wishes to specify a degree regardless of RSQ (by setting MD to a desired value), then the user should set RSQ = 100.

Array S, of dimension MD + 3, contains statistical data of interest upon return. S (1) contains the sum of squares of the y's. S (2) is the sum of squares attributable to the mean, or N times the square of the mean of y. When S (2) is subtracted from S (1), the result is the value of

$$\sum_{i=1}^{N} (y_i - y)^2$$

The next ID entries are the sums of squares attributable to each of the polynomials of degree less than or equal to ID. The last entry, in S (ID + 3), is the sum of squares of error for the entire polynomial fit. If this number is divided by N − ID − 1 and then its square root is taken, the result is the coefficient of determination. Other results can be obtained, such as obtaining the sum of squares of the errors for a polynomial of degree ID − 1 by adding S (ID + 3) and S (ID + 2), and so on.

IER is an integer return variable with several fatal-error, or warning, values:

IER = 0	Normal return.
IER = 129	Not enough data points, or MD < 1.
IER = 130	X vector is constant.
IER = 131	RSQ was set $\leq$ 0.
IER = 68	RSQ specified incorrectly; if RSQ > 100, then RSQ is set to 100 (warning).
IER = 37	Degree > 10. RLFOTH/RLFOTW not tested for degree over 10 (warning).

The call to RLFOTW is

CALL RLFOTW (X, Y, N, RSQ, MD, W, ID, P, C, S, A, B, IER)

where all of the parameters except W are exactly the same as for RLFOTH. In this case, W is a real array of length at least N containing the weights associated with the data points stored in X and Y, ordered in the same way. If all entries in W are equal, the results from RLFOTW should be identical with those from RLFOTH. Upon return, W contains the square roots of the input weights.

Two subsidiary programs of interest for users of RLFOTH and RLFOTW are RLDOPM and RLOPDC. RLDOPM converts the polynomial fit found by either RLFOTH or RLFOTW from orthogonal polynomial form, that is, from the form

$$y = c_0 P_0(x) + c_1 P_1(x) + \cdots + c_n P_n(x)$$

to the power-of-x form, that is, to the form

$$y = a_0 + a_1 x + a_2 x^2 + \cdots + a_n x^n$$

The call to RLDOPM is

    CALL RLDOPM (C, ID, A, B, T)

where C is the array found by RLFOTH or RLFOTW, ID is the degree of the polynomial, A and B are the arrays of α's and β's found by RLFOTH or RLFOTW, and T is a work array of length ID + 3 or greater. The output from RLDOPM is returned in array C, which now contains the constant term and the coefficients of the powers of x. The scaling constants that were in C(ID + 2) and C(ID + 3) remain in those positions. They are needed by RLDOPM in making the conversion to single-polynomial form because the new polynomial is unscaled in x (i.e., x is no longer scaled to the range $[-2, 2]$).

Routine RLOPDC evaluates the least-squares polynomial expression produced by either RLFOTH or RLFOTW. The subroutine operates on an input array of N independent variable values (x's) to produce N corresponding values of y. The call to RLOPDC is

    CALL RLOPDC (X, N, A, B, C, ID, IOPT, P, YHAT, IER)

where A, B, and C are the arrays from RLFOTH or RLFOTW, P is a real, double-precision work array of length at least 2N, N is the number of abscissas stored in the input array X, and YHAT is the real array of the N values of the polynomial expression on output. If the x's are scaled to the range $[-2, 2]$, then IOPT should be set to 1. If the x's are unscaled and lie in the same range as the original data values, then IOPT should be set to 2. ID is the degree of the polynomial expression. The normal output value of IER is zero. A warning of IER = 33 is issued if ID > 10.

IMSL Cubic Spline Routines ICSICU **and** ICSEVU   FORTRAN subroutine ICSICU computes the spline coefficients to interpolate a cubic spline through a set of points furnished by the user. These coefficients, together with the abscissas of the points at which the spline is to be evaluated, can be furnished to the subroutine ICSEVU, which will return the interpolated ordinates.

The call to ICSICU is

    CALL ICSICU (X, Y, NX, BPAR, C, IC, IER)

where X and Y are real arrays, with dimension of at least NX, containing the abscissas and ordinates of the NX knots. The contents of X must be ordered: X(1)

$< X(2) < X(3) \cdots$. The coefficients are returned in the real array C of dimension NX - 1 by 3. IC is the row dimension of C in the calling program. (Remember that FORTRAN stores two-dimensional arrays by slicing the array up by columns and storing the columns end to end. The subroutine needs to know how long those columns are, in case the array C had dimension greater than NX - 1.) IER, the error flag on output, has the following possibilities:

IER = 0	Normal termination
IER = 129	IC < NX - 1
IER = 130	NX < 2
IER = 131	Input x's not ordered properly

The input parameter BPAR is the name of a real array of length 4 used for control of the end conditions. The values in BPAR must satisfy the conditions

$$2.0*F_1'' + BPAR(1)*F_2'' = BPAR(2)$$
$$BPAR(3)*F_{NX-1}'' + 2.0*F_{NX}'' = BPAR(4)$$

where F_i'' is the second derivative at $x = x_i$. If specific values are desired for the second derivatives at the endpoints, then BPAR should be set

BPAR(1)	=	0
BPAR(2)	=	$2.0*F_1''$
BPAR(3)	=	0
BPAR(4)	=	$2.0*F_{NX}''$

which means that the natural spline condition can be set by making all entries in BPAR equal to zero. The IMSL manual suggests other possibilities as well.

The call to ICSEVU is

CALL ICSEVU (X, Y, NX, C, IC, U, S, M, IER)

where X, Y, NX, C, and IC are the same as in ICSICU. M is the number of elements in the real array U that contains the abscissas of the points at which evaluation of the spline is desired. Real array S, of length at least M, contains the ordinates of the interpolated points upon return. The error flag IER is for warnings only. If some U(I) is less than X(1), then IER = 33. However, if some U(I) is greater than X(NX), then IER is set to 34.

6.5.6 Other Kinds of Equations

Many other kinds of curves can be adjusted by means of least squares to fit sets of data. Many of these require ingenuity in their use and more often than not require iterative solution of the normal equations because these are no longer linear. In this section we look mainly at methods of transformation from one form of the problem to another in which the simple least-squares procedures above can be used. It is necessary to recognize at the outset that the solutions found this way will differ to some extent from those obtained by brute-force use of nonlinear solution methods to minimize the sum of squares directly. Sometimes this can be overcome by the clever use of weights.

A common kind of function is the exponential function, which may be required to be fit to growth or decay curves for theoretical reasons. The form is $y = ae^{bx}$, where a and b are to be determined from the data. The trick here is to use logarithms on the equation in order to produce the equation

$$\ln y = \ln a + bx$$

If we let $Y_i = \ln y_i$ and let $A = \ln a$, then we have

$$Y = A + bx$$

which is an ordinary linear regression problem. Once we have found A and b, we can write the final equation as

$$y = e^A e^{bx} = e^{A+bx}$$

Example 6.7

Table 6.8 shows a set of data in which the values of $2e^{-0.5x}$ have had random numbers added that came from a distribution uniform on $[-0.25, 0.25]$. The values of $\{y_i\}$ were converted to the set $\{Y_i\}$ by means of the transformation $Y = \ln y$ for use in the linear regression program. The solution of the normal equations yielded $A = 0.7912316$ and $b = -0.5699830$, with a standard error of estimate of 0.1410. The value of A corresponds to $a = 2.2061$, so that

$$y = 2.2061e^{-0.56998x}$$

is the best-fitting exponential curve by least squares.   □

Table 6.8   Random Data to Be Fit with an Exponential Function by Least Squares

I	x_i	y_i
1	−1.0	3.476218
2	−0.5	2.738829
3	0.0	2.120536
4	0.5	1.807425
5	1.0	1.374689
6	1.5	0.9778061
7	2.0	0.8280637
8	2.5	0.7198425
9	3.0	0.4799087
10	3.5	0.5812692
11	4.0	0.3255444
12	4.5	0.2881004
13	5.0	0.2223901

Another kind of curve is a hyperbolic function of the form

$$y = \frac{a}{b + cx}$$

This can be rewritten as

$$\frac{1}{y} = \frac{b}{a} + \left(\frac{c}{a}\right)x$$

which is of the form

$$Y = B + Cx$$

where $Y = 1/y$, $B = b/a$, and $C = c/a$. All of the Y_i are reciprocals of the y_i. (y should never be zero in this model.) Because $b = aB$ and $c = aC$, the final result is

$$y = \frac{1}{B + Cx}$$

where B and C can be determined by linear regression.

Other kinds of functions lead to nonlinear systems to be solved. For example, a mathematical model $y = a + b \ln x$ can be treated as a linear model if we let $X_i = \ln x_i$. However, the model

$$y = a + b \ln (x + c)$$

cannot be solved without resorting to methods for solving systems of nonlinear equations. The set of normal equations is

$$ma + b\sum_i \ln (x_i + c) = \sum_i y_i$$

$$a\sum_i \ln (x_i + c) + b\sum_i [\ln (x_i + c)]^2 = \sum_i y_i \ln (x_i + c)$$

$$a\sum_i \frac{1}{x_i + c} + b\sum_i \frac{\ln (x_i + c)}{x_i + c} = \sum_i \frac{y_i}{x_i + c}$$

These equations can be solved by making an initial guess at the value of one of the variables and solving any two of the set for the third variable. In this case the solution might be found by guessing at c and then solving the last two equations for a and b. Then the first equation can be solved for c by recognizing that

$$\sum_i^m \ln (x_i + c) = \ln [\prod_i (x_i + c)]$$

so that

$$\prod_i (x_i + c) = e^{(\Sigma y_i - ma)/b}$$

which leads to an nth-degree polynomial equation to be solved for c. Once the new c has been determined, the cycle can be repeated.

These are examples of a few of the kinds of computational problems involved in constructing mathematical models. The computer is a powerful tool in exploring solutions to such problems, but we should never overlook human ingenuity in simplifying the mathematics before we start computing.

Exercises

1. Show that the nth-order differences of any nth-degree polynomial are constant.

2. Show that $f[x_1, x_2] = f[x_2, x_1]$.

3. Show that the symbols (the x's) in any nth-order divided difference can be permuted in any order without changing the value of the divided difference.

4. Two cubic spline pieces join at a knot. Show that if the conditions of equal slopes and equal second derivatives are imposed on both segments at the knot, then the two segments will have equal curvatures at the knot. (*Hint:* Look up the formula for curvature in any calculus text.)

5. Show by a geometrical construction that each straight-line segment in a linear spline has the equation

$$y = \alpha f(x_{k+1}) + (1 - \alpha)f(x_k)$$

where α is a *local variable*, which ranges from 0 to 1 over the interval (x_k, x_{k+1}).

6. Derive the expressions for α_k and β_k in the equations on page 158. Use the recursion formula as is and multiply by $P_k(x_i)$. Then sum over i and use the orthogonality property of the polynomials as well as the definition of γ_k to find α_k. To find β_k, multiply by $P_{k-1}(x_i)$ and sum over i. To find another expression for the resulting sum involving x_i, P_k, and P_{k-1},

$$\sum_{i=1}^{n} x_i P_{k-1}(x_i)P_k(x_i)$$

multiply the recursion relation by $P_{k+1}(x_i)$, sum over i, and then reindex the terms as necessary.

7. **(a)** For the polynomial

$$p(x) = a_0 + a_1(x - x_1) + a_2(x - x_1)(x - x_2)$$

$$+ \cdots + a_5(x - x_1)(x - x_2)\cdots(x - x_5)$$

   count the number of additions, subtractions, and multiplications necessary to evaluate $p(x)$ as it is written.

   (b) Convert the polynomial to nested form

$$p(x) = ((\cdots + a_2)(x - x_2) + a_1)(x - x_1) + a_0$$

   and count the number of operations required.

   (c) Generalize this to an nth-degree polynomial. If each kind of arithmetic operation takes about the same amount of time, what is the ratio of the time to evaluate each form of the polynomial for large n?

8. The following table has a mistake in one of the functional entries. Using the material in the text about growth of error in differences, find the error and correct it as best you can.

x	$f(x)$
0.50	0.52110
0.55	0.57815
0.60	0.63665
0.65	0.69675
0.70	0.75858
0.75	0.82232
0.80	0.88816
0.85	0.95612
0.90	1.02652
0.95	1.09948
1.00	1.17520
1.05	1.25386
1.10	1.33565

9. Find the following function values with the aid of Table 6.3, appropriate interpolation procedures, and trigonometric identities.
   (a) cos 40°40′
   (b) tan 39°25′
   (c) sin 50°32′
   (d) cot 51°10′

10. The following table gives the winners and their average speeds for the Indianapolis 500 automobile race from 1946 (when the Indy 500 resumed after World War II) through 1972.
   (a) Construct a linear regression of speed on years to show the trend of the speed during those years.
   (b) Use the linear regression line to predict the average speed of the winners during 1973–1986. Check the results with the race results (see a recent *World Almanac* or similar publication).

Year	Winner	Speed (m/h)
1946	George Robson	114.820
1947	Mauri Rose	116.338
1948	Mauri Rose	119.814
1949	Bill Holland	121.327
1950	Johnnie Parsons	124.002
1951	Lee Wallard	126.244
1952	Troy Ruttman	128.922
1953	Bill Vukovich	128.740
1954	Bill Vukovich	130.840
1955	Bob Sweikert	128.209
1956	Pat Flaherty	128.490
1957	Sam Hanks	135.601
1958	Jimmy Bryan	133.791
1959	Rodger Ward	135.857
1960	Jim Rathmann	138.767

1961	A. J. Foyt	139.130
1962	Rodger Ward	140.293
1963	Parnelli Jones	143.137
1964	A. J. Foyt	147.350
1965	Jim Clark	151.388
1966	Graham Hill	144.317
1967	A. J. Foyt	151.207
1968	Bobby Unser	152.882
1969	Mario Andretti	156.867
1970	Al Unser	155.749
1971	Al Unser	157.735
1972	Mark Donohue	163.465

11. The density of air in pounds mass per cubic foot at standard pressure versus temperature is shown in the table.

    (a) Use a cubic spline to represent the density over the temperature range. Plot the cubic spline approximation and show the data points on this graph.

    (b) Use a least-squares best-fit polynomial of low degree to represent the data. Plot this approximation and the data points.

    (c) Use a least-squares best fit of an exponential function to the data. Plot the approximation.

    (d) Compare the approximations and decide which seems to give the best fit.

Temperature (°F)	Density (lbm/ft^3)
0	0.086
32	0.081
100	0.071
200	0.060
300	0.052
400	0.046
500	0.0412
600	0.0373
700	0.0341
800	0.0314
900	0.0291
1000	0.0271

12. The index of refraction n (ratio of the speed of light in vacuum to the speed of light in the material) for fused quartz glass for various wavelengths of light, in microns, is given in the table. There are no tabulated values for 0.75 and 0.85 μ.

    Produce a cubic spline interpolation to cover the range of wavelengths shown. Check your results with the values of:

    $n(0.63) = 1.457156$

    $n(0.53) = 1.460863$

    $n(0.43) = 1.467273$

Wavelength (μ)	Index of Refraction
0.35000. . .	1.47701
0.40	1.470208

Wavelength (μ)	Index of Refraction
0.45	1.465642
0.50	1.462394
0.55	1.459973
0.60	1.458096
0.65	1.456591
0.70	1.455347
0.80	1.453371
0.90000 . . .	1.451808

13. One of the methods used by petroleum engineers to estimate the future life of an oil well or reservoir is called *decline curve analysis*. In this method the production (barrels per day, month, or year) is plotted against time, and the curve is extrapolated. There are three more-or-less standard curves used.

$$q = q_i e^{-at} \qquad \text{(constant percentage decline)}$$

$$q = \frac{q_i}{1 + at} \qquad \text{(harmonic decline)}$$

$$q = \frac{q_i}{(1 + nat)^{1/n}} \qquad \text{(hyperbolic decline)}$$

where q_i is the initial rate of production, q is the rate at any later time, and a and n are constants. The choice of decline curve model depends upon the kind of well (e.g., gravity drainage wells tend to fit the harmonic decline model the best; see Slider [1976]).

The following data are for a particular well:

Month	Production (barrels/day)
1	258
2	210
3	208
4	182
5	185
6	170
7	152
8	130
9	131
10	125
11	109
12	102
13	98
14	95
15	89
16	84
17	75
18	71

(a) Fit the constant-percentage-decline model to the data. Show a graph of the exponential along with the data points. (Use a logarithmic transformation.)

(b) Fit the harmonic decline model to the data. Show a graph of the harmonic decline curve along with the data points. (Use a reciprocal transformation.)

14. The profile of the intensity versus wavelength of a certain type of astronomical spectral line is known theoretically to be Gaussian; that is, the profile may be described as

$$I(x) = Ae^{-b(x - x_0)^2}$$

where A and b are constants depending upon the condition of the radiating atoms, x is the wavelength in angstroms of the observed radiation, x_0 is the wavelength of the center of the line, and I is the intensity in arbitrary units. The following table contains measurements of the intensity of the line and the wavelength deviations from an arbitrary zero setting of the measuring instrument (near the center of the line). Use the method of least squares to find values for A, b, and x_0 (use a logarithmic transformation). Show the line profile on a graph along with the experimental points.

Wavelength Deviation (Å)	Intensity (arbitrary units)
−2.305	0.0604
−1.839	0.3487
−1.374	1.2137
−0.908	2.5707
−0.443	3.3012
0.023	2.5691
0.691	1.2201
0.954	0.3458
1.419	0.0701

15. An oil well was pumped at varying rates in a pressure-production test. It was pumped for a long time at a relatively low rate, and then at $t = 0$ the rate was stepped up to five times the low rate for an hour. During the following hour it was pumped at the low rate again; then for another hour it was pumped at four times the low rate; and for the last hour it was pumped at twice the low rate. During this time the reservoir pressure was monitored at a certain point, with results shown in the accompanying table. A plot of the data indicates that "sharp corners" on the graph are a real phenomenon.

(a) Produce a graph of the data using a spline interpolator.
(b) Modify the procedure so that four segments of the graph are drawn from four different splines, which intersect at the corner points. Compare with the graph drawn in (a).

Time (h:min)	Pressure (psia)
−0:15	2000
0:00	2000
0:15	1395
0:30	1345
0:45	1310
1:00	1290
1:15	1780
1:30	Missing
1:45	1805
2:00	1810

Time (h:min)	Pressure (psia)
2:15	1500
2:30	1480
2:45	1470
3:00	1460
3:15	1610
3:30	1620
3:45	1625
4:00	1630

16. For the following sets of data, find the best-fitting least-squares polynomial. Choose the degree of the polynomial by watching the size of the standard error of estimate as the degree is increased until the standard error of estimate levels off.

	x	y
(a)	-1.0	4.025
	-0.5	3.159
	0.0	3.526
	0.5	3.024
	1.0	2.909
	1.5	1.571
	2.0	-0.828
	2.5	-4.267
(b)	1.01	1.651
	1.50	2.352
	1.99	2.898
	2.51	3.592
	3.02	3.437
	3.48	3.351
	4.01	3.106
	4.50	2.325
(c)	-1.0	4.662
	-0.5	2.938
	0.0	2.089
	0.5	1.774
	1.0	1.770
	1.5	1.974
	2.0	2.299
	2.5	2.664

Problems

PROJECT PROBLEM 6.1

Rewrite the spline evaluation function SPLEVAL in PASCAL so that the function tests whether the argument for the present call is in the same interval as for the last call (this is very common, for example, if the spline is used for graph plotting). A suggestion is to use a global variable to record the integer identifier of the last interval called. Test this interval first upon entry. If it is not the correct interval, then institute a binary search for the correct interval. Show that the sequential search in this function requires an average of $N/2$ tests before success, and that the number of tests for binary search is proportional to $\log_2 N$. Compare the values of these two functions of N for $N = 10$, 100, and 1000.

PROJECT PROBLEM 6.2

The U.S. Standard Atmosphere (1962) is a table of values of temperature, pressure, and atmospheric density versus altitude in feet above the surface of the earth. These numbers represent averages, and there may be substantial departures from these values at various times.

Construct spline approximations for some of the quantities, including the density ratio (ratio of density at altitude to the mean sea-level air density). The density ratio function will be used in one of the projects in Chapter 8.

Altitude (ft)	Temperature (°F)	Pressure Ratio (p/p_0)	Density Ratio (ρ/ρ_0)
0	59.00	1.0000	1.0000
5,000	41.17	0.8320	0.8617
10,000	23.34	0.6877	0.7385
15,000	5.51	0.5643	0.6292
20,000	−24.62	0.4595	0.5328
25,000	−30.15	0.3711	0.4481
30,000	−47.99	0.2970	0.3741
35,000	−65.82	0.2353	0.3099
40,000	−69.70	0.1851	0.2462
45,000	−69.70	0.1455	0.1936
50,000	−69.70	0.1145	0.1522
55,000	−69.70	0.09001	0.1197
60,000	−69.70	0.07078	0.09414
65,000	−69.70	0.05566	0.07403
70,000	−67.30	0.04380	0.05789
75,000	−64.55	0.03452	0.04532
80,000	−61.81	0.02725	0.03553
85,000	−59.07	0.02155	0.02790
90,000	−56.32	0.01707	0.02195
95,000	−53.58	0.01354	0.01730
100,000	−50.84	0.01076	0.01365

Make graphs of the functions, showing the interpolated points. Use enough plotting points to make a smooth graph. (Use a computer plotting routine, if possible.)

PROJECT PROBLEM 6.3

The U.S. population for the last few decades is given in the following table.

1. Construct a natural spline to interpolate the data. If this spline is extrapolated, what does it show for the 1990 population?
2. Modify the natural spline to use the differences of the first few values and the last few values to estimate the second derivatives at the endpoints. Use this spline to extrapolate the population to 1990.
3. Use a (a) first-degree, (b) second-degree, (c) third-degree, and (d) fourth-degree polynomial least-squares curve fit to produce an approximation to the data. Use these polynomials to estimate the 1990 population.
4. Compare the results. Graphs are very helpful in visualizing the data.
5. Wait around until 1990 to see what the census shows.

Year	Population
1880	50,155,783
1890	62,947,714
1900	75,994,575
1910	91,972,266
1920	105,710,620
1930	122,775,046
1940	131,669,275
1950	150,697,361
1960	179,323,175
1970	203,235,298
1980	226,504,825

PROJECT PROBLEM 6.4

Modify CUBIC_SPLINE from natural spline conditions to conditions on the second derivative. Estimate the second derivative at each end of the splines, using divided differences, and use these estimates for z_0 and z_n.

Repeat the interpolation of the witch curve

$$y = \frac{1}{1 + x^2}$$

using this modification, and compare the results with those obtained with the natural spline interpolation. Explain why there are differences in the interior values when we have changed only the endpoint values.

You may need to print out tables of values to several significant digits in order to see the changes from the natural spline interpolation unless you have a very good graphical system.

PROJECT PROBLEM 6.5

When using linear regression we often speak about a statistical quantity called the *correlation coefficient*. This number measures the predictability of y, given a value of x. For instance, if all the data points lie on a straight line, then the value of y is exactly predicted by the value of x, and we say that x and y are *highly* (in this case perfectly) correlated. However, if all the data points lie in a circular cloud, then the knowledge of the value of x does little good in predicting the value of y and we say that the variables are uncorrelated.

The correlation coefficient r, which is a measure of the degree of correlation, is given by

$$r = \frac{n\Sigma x_i y_i - (\Sigma x_i)(\Sigma y_i)}{\sqrt{n\Sigma x_i^2 - (\Sigma x_i)^2}\sqrt{n\Sigma y_i^2 - (\Sigma y_i)^2}}$$

where (x_i, y_i) are the data points, of which there are n.

It can be shown that r has a value between +1 and −1. Positive values mean that large (small) y values are associated with large (small) x values, whereas a negative value of r means that large (small) y values are associated with small (large) x values. Complete predictability means perfect correlation, or values of $r = +1$ or $r = -1$. Values of r near 0 mean that little information about the value of y is given by a value of x; random samples drawn from populations of uncorrelated variables frequently show values of r as large in magnitude as 0.2, so we should not regard r's of this size as significant.

1. Show that if $y = ax$, then $r = +1$ or $r = -1$, depending upon the sign of a. (Remember that $(a^2)^{1/2} = |a|$.)
2. Write a linear regression program that allows keyboard (and/or disk file) input of data. Have the program calculate the coefficients of the regression equation, the standard error of estimate, and the correlation coefficient.
3. Modify the program to allow keyboard input of values of x for which values of y are desired.
4. Show that the sum of squares of the deviations (SSD) can be calculated from

$$\text{SSD} = \sum_{i-1}^{n} y_i^2 - a\sum_{i=1}^{n} y_i - b\sum_{i=1}^{n} x_i y_i$$

where a and b are the coefficients of the regression line equation $y = a + bx$, so that the standard error of estimate can be calculated from

$$\left(\frac{\text{SSD}}{n - 2}\right)^{1/2}$$

[Guttman and Wilks, 1965].

5. Test your program on the following sets of data:

	x	y
Set 1:	1.21	2.315
	6.39	10.085
	−0.56	−0.340
	3.15	5.225
	2.73	4.595
	7.01	11.015
Set 2:	2.0	2.1163
	2.5	1.1264
	3.0	0.8998
	3.5	0.1984
	4.0	−0.4381
	4.5	−0.6137
Set 3:	−0.381	−4.614
	1.998	−3.535
	−1.236	1.924
	0.484	3.153
	−1.022	−0.716
	3.363	−0.159
	2.341	3.163
	−1.135	6.049
	4.012	1.902

The approximate values of r for the three sets are 1.0, −0.98, and −0.08.

PROJECT PROBLEM 6.6

1. Use your imagination and make a freehand drawing of some function on graph paper. Make it wiggle up and down two or three times.
2. Carefully measure the coordinates of a dozen or so points on your graph.
3. Use (a) a cubic spline and (b) a polynomial interpolator to fit curves through the points you chose. Find the coordinates of several dozen interpolated points.
4. Plot the original and the interpolated points on a fresh sheet of graph paper. Construct a smooth curve (use French curves, if possible, or a drafting spline) through the points.
5. Compare the new curve with the original.

PROJECT PROBLEM 6.7

1. Write out the normal equations for a linear regression equation $y = a + bx$ for two data points, (x_1, y_1) and (x_2, y_2), and solve for a and b. Show that this is the same solution as the one obtained by solving for a and b with a linear interpolating polynomial:

$$y_1 = a + bx_1$$
$$y_2 = a + bx_2$$

Thus for the linear regression case, show that the use of only two data points reduces the linear regression to a linear interpolation.

2. Write out the normal equations for a linear regression of the form $y = a + bx$ for a single data point (x_1, y_1). Show that the system has dependent equations and that the system has an infinity of solutions passing through (x_1, y_1).

3. On the basis of this exercise and your knowledge of related mathematics, generalize these results to the case of a polynomial regression of degree n.

PROJECT PROBLEM 6.8

Program a function ORTHOPOLYVAL that will evaluate, for a given k and x, the polynomial sum

$$a_0 p_0(x) + a_1 p_1(x) + \cdots + a_k p_k(x)$$

where the $p_j(x)$ are orthogonal polynomials over the data set $\{x_i\}$ and the a_j are the coefficients determined by FIND_COEFFS. Inputs to the function will include the vectors ALPHA and BETA, the parameter K (maximum degree), and the independent variable X at which the polynomial is to be evaluated.

The function P can be converted to a procedure that evaluates the orthogonal polynomials recursively, as before. However, upon return to each level of recursion, the procedure should store the value of $p_k(x)$ in the kth element of an array to be returned to ORTHOPOLYVAL. Thus we avoid repeatedly evaluating the low-degree polynomials.

Listings

LISTING OF FUNCTION LAGRANGE
(Lagrange polynomial interpolator)

```
FUNCTION LAGRANGE (X:     VECTOR;
                   Y:     VECTOR;
                   N:     INTEGER;
                   V:     REAL) :    REAL;
```

```
{Function LAGRANGE performs the evaluation of an
  interpolating polynomial as the sum of Lagrange polynomial
  terms times values of y.   X is a vector of x values and Y
  is a vector of the corresponding ordinates.   N is the
```

number of points. U is the abscissa at which the
interpolating polynomial is to be evaluated. VECTOR type
must be declared in the calling program as a one-
dimensional ARRAY of sufficient length to accommodate the
data. The calling program should make certain that the x's
are distinct.}

```
VAR

   TERM:          REAL;
    I,J:          INTEGER;
    SUM:          REAL;

BEGIN

  SUM := 0.0;
  FOR I := 1 TO N DO
    BEGIN
      TERM := 1.0;
      FOR J := 1 TO N DO
        IF I <> J THEN
           TERM := TERM*(V - X[J])/(X[I] - X[J]);
      SUM := SUM + TERM*Y[I];
    END;
  LAGRANGE := SUM;

END;
```

LISTING OF PROCEDURE POLYCOEFF

(Produces coefficients for nested divided-difference polynomial interpolator)

```
PROCEDURE POLYCOEFF (N:          INTEGER;
                     X:          VECTOR;
                VAR  Y:          VECTOR);
```

{Procedure POLYCOEFF provides the set of coefficients for
the divided difference nested interpolating polynomial.

POLYCOEFF should be called with the following parameters:

N -   The number of base points stored in vector X

X -   The abscissas of the data points

Y -   The ordinates of the data points (input)
         Set of polynomial coefficients (output)

At the end of POLYCOEFF the ordinates (Y) will be replaced
by the set of coefficients of the interpolating polynomial.

The calling program MUST include a type declaration for
type VECTOR:

```
     TYPE        VECTOR = ARRAY[0..50] OF REAL; }



VAR
          I,J:        INTEGER;


BEGIN

  FOR I := 1 TO N-1 DO
    BEGIN
      FOR J:= 1 TO N-I DO
        BEGIN
          Y[N-J+1] := (Y[N-J+1] - Y[N-J])/
                      (X[N-J+1] - X[N-I-J+1]);
        END;
    END;

END;
```

LISTING OF FUNCTION POLYVAL

(Nested divided-difference polynomial interpolator)

```
FUNCTION POLYVAL (N:        INTEGER;
                  X:        VECTOR;
                  Y:        VECTOR;
                  V:        REAL):      REAL;



{Function POLYVAL evaluates a divided-difference
 interpolating polynomial whose coefficients are produced
 by the procedure POLYCOEFF.  The polynomial evaluation is
 performed in nested fashion as described in the text.

 Inputs are:

 N -      The number of points used in the interpolation

 X -      Vector of base points for the interpolation
```

Y -   Vector of coefficients of the nested polynomial
           produced by POLYCOEFF

V -   The value of x at which the polynomial is to be
           interpolated

POLYCOEFF must be used before POLYVAL in order to produce
the set of coefficients Y. }


```
VAR     I:    INTEGER;
        T:    REAL;

BEGIN

  T := Y[N];
  FOR I := N-1 DOWNTO 1 DO
    BEGIN
      T := T*(V - X[I]) + Y[I];
    END;
  POLYVAL := T;

END;
```


LISTING OF PROCEDURE CUBIC_SPLINE

(Produces coefficients for cubic spline interpolator)

```
PROCEDURE CUBIC_SPLINE (N:           INTEGER;
                        B:           VECTOR;
                        C:           VECTOR;
                        D:           VECTOR;
                        Y:           VECTOR;
                 VAR    Z:           VECTOR) ;
```


{This procedure evaluates the coefficients for a cubic
spline interpolator. See the text for details of the
process.

Parameters are as follows:

N - The number of knots. There are N-1 spline pieces
     which join at the knots. The slopes and the second
     derivatives (curvatures) of the joining pieces are
     equal at each knot.

B - Vector of abscissas of the N knots.

Y - Vector of ordinates of the N knots.

C,D - Vectors of constants generated by CUBIC_SPLINE
      which need to be sent to the tridiagonal equation
      solver as input to find vector Z.

Z - Vector of coefficients needed by the spline evaluation
     function SPLEVAL. Solved for in TRIDIAG.

Note that the tridiagonal equation solver TRIDIAG is
included and called from this procedure.

Also note that the calling program MUST declare a type
VECTOR as a one-dimensional array:

```
   TYPE        VECTOR = ARRAY[1..50] OF REAL;   }



VAR      I:    INTEGER;
         T:    REAL;

PROCEDURE TRIDIAG(N: INTEGER; VAR A,B,C,D: VECTOR);

VAR   MULT:   REAL;
         I:   INTEGER;

BEGIN

  FOR I := 2 TO N DO
    BEGIN
      MULT   := A[I-1]/D[I-1];
      D[I]   := D[I] - MULT*C[I-1];
      B[I]   := B[I] - MULT*B[I-1];
    END;

    B[N]  := B[N]/D[N];

  FOR I := N-1 DOWNTO 1 DO
              B[I]  := (B[I] - C[I]*B[I+1])/D[I];

END;   {TRIDIAG}


BEGIN   {Procedure CUBIC_SPLINE}

C[1] := 0.0; D[1] := 1.0; Z[1] := 0.0;
FOR I := 2 TO N-1 DO
  BEGIN
    D[I] := 2.0*(B[I+1] - B[I-1]);
    C[I] := B[I+1] - B[I];
    T    := (Y[I+1] - Y[I])/C[I];
    Z[I] := 6.0*(T - (Y[I] - Y[I-1])/(B[I] - B[I-1]));
  END;
```

```
Z[N]   : =  0.0;
C[N-1] : =  0.0;
D[N]   : =  1.0;

TRIDIAG(N,C,Z,C,D);

END;
```

LISTING OF FUNCTION SPLEVAL

(Spline function evaluator)

```
FUNCTION  SPLEVAL (N:        INTEGER;
                   X:        VECTOR;
                   Y:        VECTOR;
                   Z:        VECTOR;
                   V:        REAL) :      REAL;
```

```
{Function SPLEVAL computes the value of the spline
 approximation for a value of the variable V.   The
 appropriate spline segment is determined by a linear
 search.   If V lies outside of the domain enclosed between
 the first and last knots, an extrapolation is made using
 the first or last spline segment as appropriate.

 Parameters:

 X -  Vector containing the abscissas of the knots.

 Y -  Vector containing the ordinates of the knots.

 Z -  Vector of constants produced by procedure
      CUBIC_SPLINE.

 N -  The number of knots.   There will be N-1 spline
      segments which join at the knots with equal slopes
      and equal second derivatives (curvatures).

 V -  The value of x at which the spline is to be evaluated.


 The calling program must declare data type VECTOR which is
 an ARRAY of REAL of sufficient length to contain the data.}
```

```
    VAR
                    I:      INTEGER;
          P, Q, H, T:      REAL;


BEGIN

I := N;
REPEAT          {Procedure to identify segment in which V lies}
  BEGIN
    I := I - 1;
    T := V - X[I];
  END
UNTIL  ((T >= 0.0)  OR  (I = 1));

H := X[I+1] - X[I];                              {Evaluate spline}
P := T*(Z[I+1] - Z[I])/(6.0*H) + Z[I]/2.0;
Q := T*P + (Y[I+1] - Y[I])/H - (H/6.0)*(2.0*Z[I] + Z[I+1]);
SPLEVAL := Q*T + Y[I];

END;
```

LISTING OF PROCEDURE ORTHOG

(Produces polynomials orthogonal over a data set)

```
PROCEDURE ORTHOG (M:         INTEGER;
                  N:         INTEGER;
                  X:         DATAVECTOR;
      VAR    ALPHA:          COEFVECTOR;
      VAR     BETA:          COEFVECTOR;
      VAR    GAMMA:          COEFVECTOR;
      VAR       PX:          MATRIX);
```

```
{This procedure produces a set of polynomials of degree
 0 <= degree <= N  that are orthogonal with respect to the
set of x's for use in least-squares curve fitting.  The
value of N must be less than M, but this is not checked in
this procedure.

The calling program must declare the following data types:

  TYPE  COEFVECTOR = ARRAY[0..M] OF REAL;

  TYPE  DATAVECTOR = ARRAY[1..N] OF REAL;

  TYPE  MATRIX = ARRAY[0..N, 1..M] OF REAL;

Upon return, the array PX is filled with the values of all
of the orthogonal polynomials at all of the data points;
```

that is, PX[J,K] contains the value of the J-th degree
polynomial at the K-th data point.}

```
VAR
  I, J:          INTEGER;
    SUM:          REAL;


BEGIN

  FOR I := 1 TO M DO    PX[0, I] := 1.0;
  SUM := 0.0;
  FOR I := 1 TO M DO    SUM := SUM + X[I];
  ALPHA[0] := SUM/M;
  GAMMA[0] := M;

  FOR J := 1 TO N DO
    BEGIN
      FOR I := 1 TO M DO
        IF J > 1 THEN
          PX[J,I] := (X[I] - ALPHA[J-1])*PX[J-1,I]
                       - BETA[J-1]*PX[J-2,I]
        ELSE
          PX[J,I] := (X[I] - ALPHA[J-1])*PX[J-1,I];
      SUM := 0;
      FOR I := 1 TO M DO    SUM := SUM + SQR(PX[J,I]);
      GAMMA[J] := SUM;
      SUM := 0;
      FOR I := 1 TO M DO    SUM := SUM + X[I]*SQR(PX[J,I]);
      ALPHA[J] := SUM/GAMMA[J];
      BETA[J] := GAMMA[J]/GAMMA[J-1];
    END;

END;
```

LISTING OF PROCEDURE FIND_COEFFS

(Finds coefficients in least-squares polynomial fit)

```
PROCEDURE  FIND_COEFFS (     M:        INTEGER;
                             N:        INTEGER;
                            PX:        MATRIX;
                         GAMMA:        COEFVECTOR;
                             Y:        DATAVECTOR;
                   VAR    A:        COEFVECTOR);
```

{Procedure FIND_COEFFS determines the coefficients A needed
to determine a least-squares fit for a set of polynomials
that are orthogonal over a finite set of data points.

The input parameters are:

  M - the number of data points

  N - the degree of the polynomial to be fitted

  PX - matrix of values of all polynomials at all data
      points

  GAMMA - array of coefficients found by ORTHOG

  Y - array of data point ordinates

  A - array of coefficients needed to form a least-squares
    polynomial fit using the set of orthogonal poly-
    nomials as a basis

The calling program must declare the following types:

```
TYPE COEFVECTOR = ARRAY[0..N] OF REAL;

TYPE DATAVECTOR = ARRAY[1..M] OF REAL;

TYPE MATRIX = ARRAY[0..N,1..M] OF REAL; }



VAR                  I,J:      INTEGER;
                     SUM:      REAL;

BEGIN


  FOR J := 0 TO N DO
    BEGIN
      SUM := 0;
      FOR I := 1 TO M DO   SUM := SUM + PX[J,I]*Y[I];
      A[J] := SUM/GAMMA[J];
    END;

END;
```

LISTING OF PROCEDURE STD_ERROR

(Calculates standard error of polynomial fit)

```
PROCEDURE STD_ERROR ( M:        INTEGER;
                      N:        INTEGER;
                      PX:       MATRIX;
```

```
                        Y:          DATAVECTOR;
                        A:          COEFVECTOR;
               VAR STERROR:         COEFVECTOR) ;
```

{The procedure STD_ERROR calculates the standard error of
the fit of the least-squares polynomial constructed from a
set of polynomials orthogonal over the abscissas stored in
vector X. The set of coefficients for the least-squares
approximation is in the vector A. The set of values of
the individual polynomials is contained in the matrix PX
used in the orthogonalization process.

The parameters are:

  M - the number of data points

  N - the degree of the polynomial

  PX - matrix of values of all polynomials at all data
       points

  Y - array of ordinates of the data points

  A - array of coefficients of the orthogonal polynomials
       used in forming the least-squares polynomial fit

  STERROR - array of standard errors of estimate

The calling program must declare the following types:

  TYPE COEFVECTOR = ARRAY[0..N] OF REAL;

  TYPE DATAVECTOR = ARRAY[1..M] OF REAL;

  TYPE MATRIX = ARRAY[0..N, 1..M] OF REAL; }

VAR
                    I, J, K: INTEGER;
               SUM, SUMK: REAL;

BEGIN

  SUM := 0;
  FOR I := 1 TO M DO SUM := SUM + SQR(Y[I] - A[0]);
  STERROR[0] := SQRT(SUM/(M-1));
```
```

```
FOR J := 1 TO N DO
  BEGIN
    SUM := 0;
    FOR I := 1 TO M DO
      BEGIN
        SUMK := 0;
        FOR K := 0 TO J DO
          SUMK := SUMK + A[K]*PX[K,I];
        SUM := SUM + SQR(Y[I] - SUMK);
      END;
    STERROR[J] := SQRT(SUM/(M-J-1));
  END;

END;
```

LISTING OF FUNCTION P

(Evaluates orthogonal polynomials recursively)

```
FUNCTION P              (K:    INTEGER;
                    ALPHA:    COEFVECTOR;
                     BETA:    COEFVECTOR;
                        V:    REAL): REAL;
```

{Function P evaluates any one of the individual members of
the set of polynomials orthogonal over a set of data
points. The evaluation is done recursively.

The parameters are:

    K - the degree of the desired polynomial

    ALPHA - array of coefficients produced by ORTHOG

    BETA - array of coefficients produced by ORTHOG

    V - value of the independent variable for which the
        polynomial is to be evaluated

The calling program must declare a type:

    TYPE COEFVECTOR = ARRAY[0..N] OF REAL;

where N is sufficiently large to accommodate the arrays
ALPHA and BETA; N must be at least as great as the degree
of the polynomials to be used.}

```
BEGIN

    IF K < 0 THEN P := 0

    ELSE IF K = 0 THEN P := 1.0

    ELSE P := (V - ALPHA[K-1])*P(K-1,ALPHA,BETA,V)

                - BETA[K-1]*P(K-2,ALPHA,BETA,V)

END;
```

7

Numerical Evaluation
of Integrals

7.1 Introduction

Science and engineering calculations often end up with an integral that needs evaluation. Even in such a simple system as a pendulum (see Figure 7.1), the calculation of the period of the pendulum leads to a definite integral requiring numerical computation if we do not make the usual small-angle approximation of $\sin \theta$ by θ. The simple pendulum has a differential equation

$$\frac{d^2\theta}{dt^2} = \left(\frac{g}{L}\right) \sin \theta$$

where θ is the instantaneous angle of the pendulum from vertical (see Figure 7.1), L is the length of the pendulum, g is the acceleration of gravity, and t is the time. If θ_0, the amplitude, is small, $\sin \theta$ can be approximated by θ so that the equation is linear and can be solved very easily, but if it is not small, then we must resort to numerical methods such as those in the next chapter for the solution of the angle versus time. Even the period T of the pendulum, given by $2\pi\sqrt{(L/g)}$ in the small-angle case, requires numerical treatment in the nonlinear case.

Let $d^2\theta/dt^2 = d\dot{\theta}/dt$, where $\dot{\theta} = d\theta/dt$, and then let $d\dot{\theta}/dt = (d\dot{\theta}/d\theta)(d\theta/dt) = \dot{\theta}\, d\dot{\theta}/d\theta$. Using this substitution we can separate variables and solve to find

$$\left(\frac{1}{2}\right)\dot{\theta}^2 = C - \left(\frac{g}{L}\right)\cos \theta$$

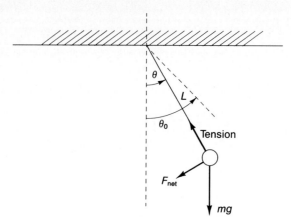

FIGURE 7.1   Simple pendulum.

Since $\dot{\theta} = 0$ when $\theta = \theta_0$, we can eventually find that

$$T = 4\sqrt{\frac{L}{g}} \int_0^{\theta_0} \frac{d\theta}{\sqrt{2(\cos\theta - \cos\theta_0)}}$$

We have no way to evaluate this integral except numerically. Furthermore, it has an endpoint singularity in the integrand because the denominator goes to zero at $\theta = \theta_0$. Nonetheless, it is possible to transform the integral by making the substitutions $k = \sin(\theta_0/2)$ and $\sin(\theta/2) = k \sin\phi$, so that

$$T = 4\sqrt{\frac{L}{g}} \int_0^{\phi_0} \frac{d\phi}{\sqrt{1 - k^2\sin^2\phi}}$$

where $\phi_0 = \operatorname{Sin}^{-1}[(1/k) \sin\theta_0/2]$. It is now in a form suitable for numerical integration (no singularity on the region $[0, \phi_0]$).

Some of you may recognize this integral as an elliptic integral of the first kind, for which there are extensive tables laboriously calculated by hand in days gone by.

In this chapter we will look at several methods of evaluating definite integrals numerically. Most of the methods depend on the basic idea of replacing the integrand function by some sort of interpolating polynomial, which can be integrated very easily. The methods range from simple and not very accurate methods, such as the trapezoidal rule, to an automatic integrator, QUAD, which adapts its behavior to the nature of the integrand and tries to produce a result correct to within a specified error tolerance. But first we must look at the nature of the integration process itself.

7.2   The Definite Integral

Throughout this chapter we will deal only with the Riemann integral. We define the Riemann integral as follows:

Let a and b be the limits of integration. On the interval $[a, b]$, define a set of points x_i such that

$$a = x_0 < x_1 < x_2 < \cdots < x_n = b$$

In each subinterval $[x_i, x_{i+1}]$, define a point z_i, where $x_i \leq z_i \leq x_{i+1}$. Form the sum

$$\sum_{i=0}^{n-1} f(z_i)(x_{i+1} - x_i)$$

Figure 7.2 shows such a sum geometrically. We now take the limit as $n \to \infty$ with the condition that

$$\max_i |x_{i+1} - x_i| \to 0$$

If that limit exists, independent of the way in which the x_i's and z_i's are chosen, then we define the limit as the Riemann integral of $f(x)$ over $[a, b]$ and symbolize the integral by

$$\int_a^b f(x)\, dx$$

In most calculus courses it is proved that a sufficient condition for the existence of the limit and the integral is the continuity of $f(x)$ on $[a, b]$. It is also proved

FIGURE 7.2   Rectangles forming Riemann sum.

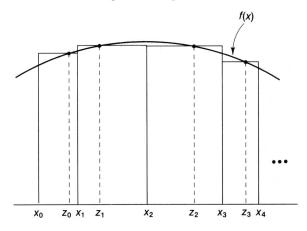

that if $f(x)$ has a finite (jump type) discontinuity in $[a, b]$ at a point c, then the integral exists by virtue of the existence of the following limit:

$$\int_a^b f(x)\,dx = \lim_{\epsilon \to 0^+} \left\{ \int_a^{c-\epsilon} f(x)\,dx + \int_{c+\epsilon}^b f(x)\,dx \right\}$$

where $\epsilon \to 0^+$ means that ϵ approaches zero from the positive side. This idea can be extended to deal with any finite number of finite discontinuities in $[a, b]$. The least-restrictive conditions for the existence of the integral are beyond the scope of this book. In fact, we will encounter numerical difficulties before we get to problems that must take account of those conditions. We will deal with some of the ways of treating improper integrals later in the chapter, but first we must look at the main methods of numerical integration.

7.3   Newton-Cotes Closed Formulas

The Newton-Cotes closed formulas include many of the familiar numerical integration methods. The term *closed* refers to the fact that the endpoints of the integration interval are included in the formulas. In the *open* methods, they are not. A characteristic of the Newton-Cotes methods is that they involve equally spaced points across the interval of integration. The starting point for these formulas is the Newton forward difference formula for interpolation:

$$f(x_0 + \alpha h) = f(x_0) + \alpha \Delta f(x_0) + \frac{\alpha(\alpha - 1)}{2!} \Delta^2 f(x_0) + \cdots$$

We regard α as the independent variable in this formula, and we recognize that if $x = x_0 + \alpha h$, then $dx = h\,d\alpha$. The first of these formulas is the trapezoidal rule, which is found by integrating $f(x_0 + \alpha h)$ from x_0 to x_1 (α from 0 to 1):

$$\int_{x_0}^{x_1} f(x)\,dx = \int_{x_0}^{x_1} f(x_0)\,dx + \Delta f(x_0) \int_{x_0}^{x_1} \alpha\,dx + \int_{x_0}^{x_1} \frac{\alpha(\alpha - 1)}{2} h^2 f''(\xi)\,dx$$

$$= hf(x_0) + h\Delta f(x_0) \int_0^1 \alpha\,d\alpha + \frac{h^3}{2} \int_0^1 f''(\xi)\alpha(\alpha - 1)\,d\alpha$$

$$= hf_0 + \frac{h(f_1 - f_0)}{2} + \frac{h^3}{2} f''(\zeta) \int_0^1 \alpha(\alpha - 1)\,d\alpha$$

$$= \frac{h(f_0 + f_1)}{2} - \frac{h^3}{12} f''(\zeta)$$

where $x_0 \le \zeta \le x_1$, where f_0 and f_1 denote $f(x_0)$ and $f(x_1)$, and we use the extended form of the mean value theorem for integrals to evaluate the integral of the error term. The error term we have developed is sometimes referred to as the *truncation error* because we have truncated an infinite series to provide an approximation to the integral.

Figure 7.3 shows a trapezoid inscribed beneath the graph of $y = f(x)$ between x_0 and x_1. Note that for the particular $f(x)$ shown, $f''(x) < 0$ on $[x_0, x_1]$. This makes

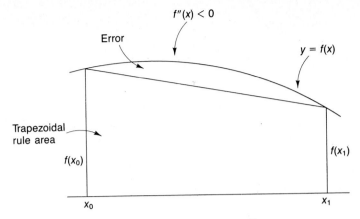

FIGURE 7.3   Single trapezoid approximating $\int_{x_0}^{x_1} f(x)\,dx.$

the error term in the trapezoidal rule formula positive in sign, which means that the area of the trapezoid is too small. The reader is encouraged to follow through the same analysis for a function $f(x)$ that has a graph that is concave upward (f'' > 0) over the interval from x_0 to x_1.

In most cases we would not find this a satisfactory method to use in estimating an integral such as

$$\int_1^3 x^x\,dx$$

because it would grossly overestimate the value of the integral (28.000000 instead of 13.725105). The trapezoidal rule, if it is used at all, is used most often in what is called a composite rule or composite formula. The interval of integration is broken up into a large number of (usually) equal subintervals, and the trapezoidal rule is used on each of the subintervals. Thus the composite trapezoidal rule is a method of summing the areas of a set of trapezoids inscribed under the graph of the function being integrated. The composite trapezoidal rule can be written as

$$\int_a^b f(x)\,dx = \left(\frac{h}{2}\right)(f_0 + f_n) + h\sum_{i=1}^{n-1} f_i$$

This will improve the accuracy, which can be seen by considering the error term for the rule: $-h^3 f''(\zeta)/12$. For a single interval $[a, b]$, the error is $-(b - a)^3 f''(\zeta)/12$, but we do not know the value of ζ or the value of $f''(\zeta)$. If we assume that $f''(x)$ is bounded in the interval, we can replace it by its largest absolute value, say M, and then divide the interval $[a, b]$ into n subintervals, each of length $(b - a)/n$. We now can calculate a bound on the error for the sum of the integrals over the n subintervals by writing

$$\left(\frac{b - a}{n}\right)^3 \left(\frac{M}{12}\right)(n) = \frac{M(b - a)^3}{12n^2}$$

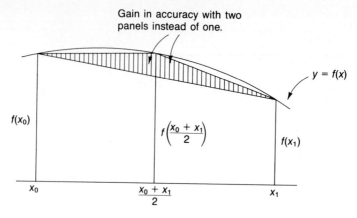

FIGURE 7.4   Two trapezoids approximating $\int_{x_0}^{x_1} f(x)\,dx$.

where the factor n in the numerator in the left-hand side of the equation accounts for the fact that there are now n such local errors. Figure 7.4 shows geometrically what the effect is when we replace a single trapezoid by two trapezoids. Although the error formula for the composite rule indicates that we should have a reduction in error by a factor of 4 every time we double the number of steps, there is a limit to the gains to be achieved by that process. There comes a time when n is so large that the accumulated round-off errors contribute more to the total error than the inaccuracy in the formula (truncation error).

Simpson's one-third rule is the next-higher Newton-Cotes closed formula in which a quadratic is used for interpolation. Thus three points are necessary in order to determine the quadratic polynomial. Simpson's rule exhibits a phenomenon of the even-ordered (even degree of the interpolating polynomial) Newton-Cotes closed formulas: Each has the same order of accuracy as the odd-ordered formula next higher; that is, the exponents on h are the same and the derivatives are of the same order, although the constants are different. As a consequence, Simpson's one-third rule not only will give exact results in the integration of any quadratic but will give exact results for cubics as well. To derive this formula we must therefore start with a Newton forward difference formula including third differences because the integral of the second difference term vanishes:

$$f(x_0 + \alpha h) = f_0 + \alpha \Delta f_0 + \frac{\alpha(\alpha - 1)}{2!}\Delta^2 f_0 + \frac{\alpha(\alpha - 1)(\alpha - 2)}{3!}\Delta^3 f_0$$
$$+ \frac{h^4 \alpha(\alpha - 1)(\alpha - 2)(\alpha - 3)}{4!}f^{(4)}(\xi)$$

where ξ is somewhere in the interval bounded by the extreme values of x, x_0, x_1, and x_2.

We integrate across two spaces, or "panels," from x_0 to x_2, using $dx = h\,d\alpha$:

$$\int_{x_0}^{x_2} f(x_0 + \alpha h)\,dx = hf_0 \int_0^2 d\alpha + h\Delta f_0 \int_0^2 \alpha^2\,d\alpha + h\frac{\Delta^2 f_0}{2!}\int_0^2 \alpha(\alpha - 1)\,d\alpha$$

$$+ h\frac{\Delta^3 f_0}{3!}\int_0^2 \alpha(\alpha - 1)(\alpha - 2)\,d\alpha + \frac{h^5}{4!}\int_0^2 f^{(4)}(\xi)\alpha(\alpha - 1)(\alpha - 2)(\alpha - 3)\,d\alpha$$

We cannot use the extended mean value theorem for integrals to find the error term because $\alpha(\alpha - 1)(\alpha - 2)(\alpha - 3)$ changes sign in the interval; instead we rely upon a result by Steffensen [1927] to help us find

$$\int_{x_0}^{x_2} f(x_0 + \alpha h)\,dx = 2hf_0 + 2h\Delta f_0 + h\frac{\Delta^2 f_0}{2} - \frac{h^5}{90}f^{(4)}(\zeta)$$

$$= \frac{h}{3}[f_0 + 4f_1 + f_2] - \frac{1}{90}h^5 f^{(4)}(\zeta), \qquad x_0 < \zeta < x_2$$

This method, Simpson's first rule, is almost always used in a composite rule form. The rule requires that the number of subintervals, or panels, be even. If we let m be the number of panels, each of width h, and let $h = (b - a)/m$, then we can write

$$\int_a^b f(x)\,dx = \left(\frac{h}{3}\right) \sum_{\substack{i=0 \\ \Delta i = 2}}^{m-2} \left[f(x_i) + 4f(x_{i+1}) + f(x_{i+2})\right]$$

This is inefficient because at almost half of the interior points the function is evaluated twice. It is more efficient computationally to write

$$\int_a^b f(x)\,dx = \frac{b - a}{3m}\left[f_0 + f_m + 4\sum_{\substack{i=1 \\ \Delta i = 2}}^{m-1} f_i + 2\sum_{\substack{i=2 \\ \Delta i = 2}}^{m-2} f_i\right]$$

The Simpson's rule error term shows that if we reduce the size of h by a factor of 2, we should expect a reduction in the local error by a factor of approximately 32, but because there are now twice as many local errors, we gain only a factor of 16 in the overall accuracy.

The next rule in the Newton-Cotes series is sometimes called the Simpson $\frac{3}{8}$ rule. The derivation of it is left as an exercise for the student. The formula is

$$\int_{x_0}^{x_3} f(x)\,dx = \left(\frac{3h}{8}\right)(f_0 + 3f_1 + 3f_2 + f_3) - \frac{3h^5}{80}f^{(4)}(\zeta), \qquad x_0 < \zeta < x_3$$

Note that the error term has the same exponent on h and the same derivative of f as in the first Simpson rule. If we make the assumption that the values of $f^{(4)}$ are approximately equal, we can compare the relative accuracy of the two methods by making the h of the $\frac{3}{8}$ rule equal to two-thirds of that of the first rule. What we find is that the $\frac{3}{8}$ rule is about twice as accurate as the second-order rule at the expense of one extra function evaluation, or about one-third more work.

Thus there is not usually a dramatic gain in going from the first to the second Simpson rule.

The next two rules are given without derivation:

$$\int_{x_0}^{x_4} f(x)\,dx = \frac{2h}{45}[7f_0 + 32f_1 + 12f_2 + 32f_3 + 7f_4] - \frac{8h^7}{945}f^{(6)}(\zeta), \qquad x_0 < \zeta < x_4$$

$$\int_{x_0}^{x_5} f(x)\,dx = \frac{5h}{288}[19f_0 + 75f_1 + 50f_2 + 75f_4 + 19f_5] - \frac{275h^7}{12096}f^{(6)}(\zeta), \qquad x_0 < \zeta < x_5$$

Note again the similarity in behavior of the error terms to that of the two Simpson rules.

Very high-order rules tend not to be used; the very high-degree integration formulas have large negative and positive coefficients, which often can lead to loss of accuracy through cancellation errors. Instead, composite rules with lower-order polynomials seem to be more satisfactory.

7.4   Newton-Cotes Open Formulas

The Newton-Cotes open formulas are derived in much the same way as the closed formulas, the difference being that the endpoints are not used and the interpo-lating polynomial is allowed to dangle at each end of the integration interval.

The first of the formulas is the rectangle, or midpoint, rule. We again use the Newton forward difference formula, this time based upon x_1:

$$f(x_1 + \alpha h) = f_1 + \alpha \Delta f_1 + \frac{\alpha(\alpha - 1)}{2!}h^2 f''(\xi)$$

where ξ is in the interval bounded by the extreme values of x, x_0, x_1, and x_2.

We integrate from x_0 to x_2:

$$\int_{x_0}^{x_2} f(x_1 + \alpha h)\,dx = h\int_{-1}^{1} f_1\,d\alpha + h\Delta f_1 \int_{-1}^{1} \alpha\,d\alpha + \frac{h^2}{2}\int_{-1}^{1} f''(\xi)\alpha(\alpha - 1)\,d\alpha$$

$$= 2hf_1 + \frac{h^3}{3}f''(\xi), \qquad x_0 < \xi < x_2$$

Note that the error term has the same exponent on h and the same order of derivative as that of the trapezoidal rule. A fair comparison requires us to rec-ognize that $2h$ is the panel width for the midpoint rule and h is the trapezoidal rule panel width. Suppose we wish to integrate from 0 to 1 with, for instance, 4 function evaluations. For the trapezoidal rule we would use $h = \frac{1}{3}$, and for the rectangle rule we would use $h = \frac{1}{8}$. Then if f'' were constant over the interval, the trapezoidal rule error magnitude would be

$$\frac{(3)\left(\frac{1}{3}\right)^3 f''}{12} = \frac{f''}{108}$$

For the rectangle rule, we would have

$$\frac{(4)\left(\frac{1}{8}\right)^3 f''}{3} = \frac{f''}{384}$$

This result may seem surprising in light of the fact that the trapezoidal rule is thought of as a more sophisticated approach than the rectangular rule.

Higher-order Newton-Cotes open formulas are

$$\int_{x_0}^{x_3} f(x)\,dx = \frac{3h}{2}[f_1 + f_2] + \frac{3h^3}{4}f''(\xi), \qquad x_0 < \xi < x_3$$

$$\int_{x_0}^{x_4} f(x)\,dx = \frac{4h}{3}[2f_1 - f_2 + 2f_3] + \frac{14h^5}{45}f^{(4)}(\xi), \qquad x_0 < \xi < x_4$$

$$\int_{x_0}^{x_5} f(x)\,dx = \frac{5h}{24}[11f_1 + f_2 + f_3 + 11f_4] + \frac{95h^5}{144}f^{(4)}(\xi), \qquad x_0 < \xi < x_5$$

Note that the same kind of adjacent-order behavior of the error terms occurs with the open formulas as did with the closed formulas.

The open formulas, when used, are generally used in composite formulas. Because the endpoints are not used, it is efficient to write program loops containing the obvious summations, such as

$$\frac{5h}{24} \sum_{\substack{i=0 \\ \Delta i = 5}}^{m-5} \left\{ 11f_{i+1} + f_{i+2} + f_{i+3} + 11f_{i+4} \right\}$$

where m is the number of panels to be used.

A word of caution is necessary concerning programming any of these formulas. It is not good practice to use

$$x_{i+1} = x_i + h$$

to calculate the set of points. If $|b - a|$ is large compared to h and h is not exact (e.g., $h = (b - a)/1000$ in a binary computer), the points x may be off by a considerable amount near b. It is better to use

$$x_i = a + \frac{i(b - a)}{m}$$

where m is the number of intervals and i and m are treated appropriately in the computing system. It is even better to let m be a power of 2 (4, 8, 16, 32, . . .) because then $(b - a)/m$ suffers no loss of accuracy in a binary computer.

7.5 Step Size and Accuracy

How do we choose the value of h, the step size, or m, the number of panels used in the numerical integration? The error term could be used if we could estimate the appropriate derivative that appears in the error term. For any integrand (other than for a few trivial ones, which should not be integrated numerically) the problem of finding the appropriate derivative and then finding its maximum value is a formidable task. Furthermore, the process overestimates the error and usually forces the use of many more steps than necessary. Nonetheless, the process would be as shown in Example 7.1.

Example 7.1

How many trapezoidal rule steps are necessary to evaluate the integral of e^{2x} from 0 to 1 with an error no greater than 10^{-6}?

The trapezoidal rule error term is $-h^3 f''(\zeta)/12$, where ζ is in the interval $[0, 1]$. Because $f'''(x) = 8e^{2x}$, which is maximum at $x = 1$, we take the maximum as $8e^2$. Also we find that $h = (b - a)/m = 1/m$. Thus we can write

$$(m)\left(\frac{1}{m}\right)^3 \left(\frac{8e^2}{12}\right) < 10^{-6}$$

or

$$m^2 > (2e^2)\left(\frac{10^6}{3}\right) = 4.926 \times 10^6$$

Any $m > 2219$ will suffice. Actually, this is an upper bound well above the necessary number of steps because near $x = 0$ the value of $f''(x)$ is close to $8e$. □

The error term can be used in a different way to estimate the error. In going from one trapezoidal step to two trapezoidal steps, there is usually a considerable reduction in the error. However, in going from 1024 steps to 2048 steps, for instance, the change is smaller. In the first case $f''(\zeta)$ is replaced by two $f''(\zeta)$ terms in which the values of ζ are widely separated and the values of f'' are likely to be quite different. In the latter case the two new ζ values for each split are very close, and the corresponding f'' values are nearly equal. Thus, in using a composite trapezoidal rule, once the step sizes become very small, doubling the number of steps reduces the total error by a factor of almost exactly 4. This information is also useful in determining the error by a method developed by Richardson [1910].

Let I^* be the true, or exact, value of the integral, and let I_m be the numerically computed value for m panels of the trapezoidal rule. Also let E_m be the true error for the computation, so that we can write

$$I^* = I_m + E_m$$

and, for $2m$ steps,

$$I^* = I_{2m} + E_{2m}$$

If $E_m \simeq 4E_{2m}$, then we can write the two equations as

$$I^* \simeq I_m + 4E_{2m}$$

$$I^* = I_{2m} + E_{2m}$$

From these we find that $E_{2m} \simeq (I_{2m} - I_m)/3$. In addition we can write

$$I^* \simeq I_{2m} + \frac{I_{2m} - I_m}{3}$$

$$\simeq \frac{4I_{2m} - I_m}{3}$$

This allows us to estimate the error in the calculation with $2m$ steps, E_{2m}, and also to make a better estimate of the value of I^* than that obtained by using I_{2m}.

Example 7.2

Table 7.1 shows the results of evaluating the integral

$$\int_0^1 e^x \, dx = 1.71828182845904 \ldots$$

using from 1 to 2048 steps of the trapezoidal rule. The first column of the table shows N, the number of steps, and the second column shows the trapezoidal rule value. Note that even with 2048 steps, the answer is not correct to the number of decimals shown. The third column shows the true error

$$E_n = e - 1 - I_n$$

for the value of N in that row. The fourth column shows the ratio of the error shown in the row above to the error in that row. The value of the ratio is very close to 4, but near the end of the table it is beginning to deteriorate from a combination of increasing round-off error and decreasing accuracy of the errors themselves. Column 5 shows the first 10 digits of the Richardson extrapolated estimate of I^* through $N = 128$, which shows as much accuracy at 128 panels as the trapezoidal rule at 2048. The last column shows the Richardson estimate of the error in the trapezoidal rule value for that row. The agreement is quite good.   □

The same kind of procedure can be done with Simpson's rule where the error-reduction factor is 16 (from the h^5 factor in the error term). In this case $E_{2m} \simeq (I_{2m} - I_m)/15$ and the estimated value of I^* is

$$I^* \simeq \frac{16 \, I_{2m} - I_m}{15}$$

The derivation of these formulas is left as an exercise.

Table 7.1

N	Trapezoidal Rule Value	Error	Error Ratio	Richardson Extrapolation	Estimated Error of I_{2N}
1	1.859140914	-1.408590858 E-01			
2	1.753931093	-3.564926400 E-02	3.95125	1.718861153	-3.506994 E-02
4	1.727221905	-8.940076094 E-03	3.98758	1.718318842	-8.903063 E-03
8	1.720518592	-2.236763698 E-03	3.99688	1.718284154	-2.234438 E-03
16	1.718841129	-5.593001097 E-04	3.99922	1.718281975	-5.591543 E-04
32	1.718421660	-1.398318382 E-04	3.99981	1.718281837	-1.398230 E-04
64	1.718316787	-3.495835153 E-05	3.99996	1.718281829	-3.495770 E-05
128	1.718290568	-8.739565601 E-06	4.00001	1.718281828	-8.73970 E-06
256	1.718284013	-2.184789992 E-06	4.00019		
512	1.718282374	-5.459824024 E-07	4.00158		
1024	1.718281965	-1.360622264 E-07	4.01274		
2048	1.718281862	-3.317472874 E-08	4.10138		

Example 7.3

Simpson's rule was used to evaluate numerically the integral

$$\int_0^1 e^x \, dx = e - 1 = 1.71828182845904\ldots$$

The numerical results were as follows.

Number of Panels	Value	Error
2	1.718861152	5.793235×10^{-4}
4	1.718318842	3.70135×10^{-5}
8	1.718284155	2.326541×10^{-6}
16	1.718281974	1.45541×10^{-7}
32	1.718281838	9.541×10^{-9}

The gain in accuracy in doubling the number of panels used can be shown by the ratios of the successive error terms: 15.6517, 15.9092, 15.9855, and 15.2543. The ratio is approaching the theoretical 16, but on the last calculation the error is so small that it is only correct to 2 or 3 significant digits because of the computer word size. It is left as an exercise to apply Richardson extrapolation to the values in this table to estimate I^* and to estimate the error for each value of N.   □

7.5.1   Simple Integrator with Error Estimate

These ideas suggest that we should never write a "simple" integrator, such as a Simpson's rule procedure, without obtaining all the information we can from the execution of it. Procedure SIMPSON is an example of what can be done in order to take advantage of that information.

SIMPSON is called from a user-written program, which sets up the parameters and displays the output. A function, called FUN in the procedure, must be supplied by the user. It must have a header

```
FUNCTION FUN(X: REAL): REAL;
```

and must assign a value to FUN. Note that the procedure requires that N be a multiple of 4. This allows the use of two Simpson's rules, one of which is twice the order of the other, so that a Richardson extrapolation can be carried out. If N were merely even, then N/2 might be odd, and Simpson's rule will not work in that situation.

The procedure calculates four interior sums and an endpoint sum so that Simpson's rules can be calculated for two different values of N with the same set of function evaluations. The call to SIMPSON is of the form

```
SIMPSON (A, B, N, I1N, I2N, ISTAR, E2N);
```

where

A	=	lower limit of integration
B	=	upper limit of integration
N	=	number of panels (must be a multiple of 4)
I1N	=	estimate by Simpson's rule using N/2 panels
I2N	=	estimate by Simpson's rule using N panels
ISTAR	=	Richardson extrapolated estimate using I1N and I2N
E2N	=	Richardson extrapolated estimate of error of I2N

The last four parameters are declared as variable parameters in the procedure and thus are used to return values.

All this information is obtained with little more work than that required to calculate I2N itself. You can look at the results of using SIMPSON on the integration of EXP (X) from 0 to 1 and compare the errors with the actual errors from the previous example. The error estimates are very favorable.

Number of Panels	I2N	E2N (estimated error)	ISTAR (estimated value)
8	1.7182841547	2.31248129 E-06	1.7182818422
16	1.7182819740	1.45376786 E-07	1.7182818287
32	1.7182818376	9.09943386 E-09	1.7182818285
64	1.7182818290	5.69586215 E-10	1.7182818284
128	1.7182818285	3.66223200 E-11	1.7182818284

These values show that the extrapolated error estimates are in reasonable agreement with the actual errors. There is no excuse for doing a Newton-Cotes numerical integration without getting an error estimate at the same time.

7.6   Romberg Integration

Suppose we start with the trapezoidal rule and compute estimates of the value of I^*, designating the results with the following notation:

$$I_{1,1} = \frac{(b - a)[f(a) + f(b)]}{2}$$

is the one-step trapezoidal rule value, and $I_{m,1}$ is the trapezoidal rule value for 2^{m-1} steps. To avoid recalculating the functional values in developing $I_{m,1}$ for successive values of m, we use

$$I_{m+1,1} = \frac{I_{m,1}}{2} + \frac{b - a}{2^m} \sum_{i=1}^{2^{m-1}} f\left[a + (2i - 1)\frac{b - a}{2^m}\right]$$

which amounts to finding the functional values only at new points. Having a set of such values, we can now calculate a set of new estimates of I^* using the Richardson formula

$$I_{m+1,2} = \frac{4I_{m+1,1} - I_{m,1}}{3}, \qquad m = 1, 2, \ldots$$

It is possible to show (exercise) that $I_{m,2}$ is actually a Simpson's rule value. Using the Richardson idea again we can write

$$I_{m+1,3} = \frac{16I_{m+2,2} - I_{m+1,2}}{15}$$

In general, we can write

$$I_{m+1,n} = \frac{4^{n-1} I_{m+2,n-1} - I_{m+1,n-1}}{4^{n-1} - 1}$$

The way that Romberg integration is usually programmed, the value of m is chosen, the calculations are done, and the array is displayed for visual inspection:

$I_{1,1}$

$I_{2,1}$ $I_{2,2}$

$I_{3,1}$ $I_{3,2}$ $I_{3,3}$

$\qquad$. $\qquad$. $\qquad$.

$\qquad$. $\qquad$. $\qquad$.

$\qquad$. $\qquad$. $\qquad$.

$I_{m,1}$ $I_{m,2}$ $I_{m,3}$ $\qquad$. $\qquad$. $\qquad$. $\qquad$ $I_{m,m}$

We observe that, as we read down the first column (the trapezoidal rule integration), the values converge toward the exact value I^*. Also, as we move down the main diagonal (for which we have $I_{k,k}$, $k = 1, 2, 3, \ldots$,) the estimate converges even more rapidly.

Table 7.2 shows the results of Romberg integration applied to the integral

Table 7.2   Romberg Integration of X^x from 1 to 3

I	$I_{m,1}$	$I_{m,2}$	$I_{m,3}$	$I_{m,4}$	$I_{m,5}$	$I_{m,6}$	$I_{m,7}$	$I_{m,8}$	$I_{m,9}$	$I_{m,10}$
1	28.000000									
2	18.000000	14.666667								
3	14.859617	13.812823	13.755900							
4	14.013401	13.731328	13.725895	13.725419						
5	13.797481	13.725508	13.725120	13.725107	13.725106					
6	13.743218	13.725131	13.725105	13.725105	13.725105	13.725105				
7	13.729635	13.725107	13.725105	13.725105	13.725105	13.725105	13.725105			
8	13.726238	13.725105	13.725105	13.725105	13.725105	13.725105	13.725105	13.725105		
9	13.725388	13.725105	13.725105	13.725105	13.725105	13.725105	13.725105	13.725105	13.725105	
10	13.725176	13.725105	13.725105	13.725105	13.725105	13.725105	13.725105	13.725105	13.725105	13.725105

$\int_1^3 x^x \, dx$. Note that 512 trapezoidal rule panels were not enough to give 8-digit accuracy, but that 128 Simpson's rule panels did suffice. Only 32 trapezoidal rule steps were necessary to produce 8 digits of accuracy for the extrapolation $I_{6,6}$.

7.7 Gauss Quadrature

All the numerical integration formulas we have seen so far are of the form

$$\sum_{i=1}^{n} w_i f(x_i)$$

which we can consider as a weighted sum of the values of the integrand function $f(x)$. The Newton-Cotes methods all employ x's that are equally spaced. Suppose we consider the x's as not being equally spaced. Instead, we use them as parameters like the w's in attempting to fit a polynomial to the integrand function in such a way as to give exact results for the highest-degree polynomial we can find. In the case of the trapezoidal rule, we can get exact results only for a first-degree polynomial (a straight line). For Simpson's first rule we saw that we can, in fact, get exact results for a cubic. Let us try to create a two-point formula, letting both the x's and the w's be parameters.

Let w_1, w_2, x_1, and x_2 be parameters and require that we get exact results for the integration of the polynomials 1, x, x^2, and x^3. Because the operation of integration is linear, we can expect exact results for any polynomial of degree up to and including 3. For reasons that will become clear later, we shall use the interval $[-1, 1]$ for the region of integration (we can always transform from any bounded region to $[-1, 1]$ with a linear change of variable). Thus we write

$$\int_{-1}^{1} 1 \, dx = 2 = w_1 + w_2$$

$$\int_{-1}^{1} x \, dx = 0 = w_1 x_1 + w_2 x_2$$

$$\int_{-1}^{1} x^2 \, dx = \tfrac{2}{3} = w_1 x_1^2 + w_2 x_2^2$$

$$\int_{-1}^{1} x^3 \, dx = 0 = w_1 x_1^3 + w_2 x_2^3$$

These equations can be solved to find $w_1 = w_2 = 1$, $x_1 = -3^{-1/2}$, and $x_2 = -x_1$.

This method gives a substantial gain in accuracy at the expense of introducing parameters with many digits. These parameters are no problem to a computer, but they were difficult to use in the pencil-and-paper days. In particular, if the integrand contained transcendental functions, the interpolation problems became unbearable and Newton-Cotes formulas with points chosen to agree with table entries were much preferred.

The restriction of the interval of integration to $[-1, 1]$ is only a minor inconvenience, and it permits the use of Legendre polynomials in solving for the x_i and w_i. Any integral over $[a, b]$ can be transformed by the substitution of

$$x = \frac{[(b - a)z + (a + b)]}{2}$$

Example 7.4

Let $x = [(3 - 1)z + (3 + 1)]/2 = z + 2$. Then when $x = 1$, $z = -1$, and when $x = 3$, $z = +1$. Therefore,

$$\int_1^3 x^x \, dx = \int_{-1}^1 (z + 2)^{z+2} \, dz \quad \square$$

The derivation of the weights w_i and the abscissas x_i could perhaps be done by the method used in the two-point case, but the solution is very tedious for higher-order formulas. Instead, we turn to the use of Legendre polynomials, which have special properties over the interval $[-1, 1]$. Although the method of solution is somewhat different, the results are exactly the same as for the simple algebraic method. Because of the involvement of the Legendre polynomials, the integration method over the interval $[-1, 1]$ is frequently referred to as Gauss-Legendre quadrature.

7.7.1 Gauss-Legendre Quadrature

We begin with the Legendre polynomials, the first few of which are

$$P_0(x) = 1$$

$$P_1(x) = x$$

$$P_2(x) = \frac{3x^2 - 1}{2}$$

$$P_3(x) = \frac{5x^3 - 3x}{2}$$

Legendre polynomials have many important properties. For example, they are related by a recursion formula

$$P_n(x) = \left(\frac{2n - 1}{n}\right) x P_{n-1}(x) - \left(\frac{n - 1}{n}\right) P_{n-2}(x)$$

and they are also the solutions of a particular ordinary differential equation that occurs in a number of places in mathematical physics.

The property of particular interest to us is that the set of polynomials is orthogonal over the interval $[-1, 1]$; that is,

$$\int_{-1}^{1} P_i(x)P_j(x)\, dx = 0 \qquad \text{for } i \neq j$$

but

$$\int_{-1}^{1} [P_i(x)]^2\, dx \neq 0$$

The integration of the product of two of the functions is analogous to the scalar, or *dot*, product of vectors, so that the idea of orthogonality of functions is a generalization of the idea of perpendicularity.

We start the general Gauss-Legendre integration formula derivation by choosing an nth-degree Lagrange polynomial to interpolate $f(x)$:

$$f(x) = \sum_{i=0}^{n} L_i(x)f(x_i) + R_n(x)$$

where

$$R_n(x) = \left\{ \prod_{i=0}^{n} (x - x_i) \right\} \frac{f^{(n+1)}(\xi)}{(n+1)!} \qquad -1 < \xi < 1$$

and the x_i are yet to be determined. We now use this in the integral of $f(x)$:

$$\int_{-1}^{1} f(x)\, dx = \int_{-1}^{1} \sum_{i=0}^{n} L_i(x)f(x_i)\, dx + \int_{-1}^{1} R_n(x)\, dx$$

$$= \sum_{i=0}^{n} f(x_i) \int_{-1}^{1} L_i(x)\, dx + \int_{-1}^{1} R_n(x)\, dx$$

and then impose the condition that if $f(x)$ is a $(2n + 1)$st-degree polynomial, the nth-degree Lagrange polynomial must give exact results for the integral. (Remember that we have $n + 1$ coefficients and n abscissas at our disposal for our nth-degree polynomial.) If the results are to be exact, we must then require that the integral of $R_n(x)$ be zero. If that happens, then we have found the expression for w_i:

$$w_i = \int_{-1}^{1} L_i(x)\, dx$$

Now if $f(x)$ is a $(2n + 1)$st-degree polynomial such that

$$f(x) = \sum_{i=0}^{n} L_i(x)f(x_i) + R_n(x)$$

it must be true that $R_n(x)$ is of degree $2n + 1$ because each of the Lagrange polynomials is only of order n. The part of R_n that is

$$\prod_{i=0}^{n} (x - x_i)$$

is of degree $n + 1$, so this means that $f^{(n+1)}(\xi)$ must be a polynomial of degree

n. (Remember that although x does not appear explicitly in $f^{(n+1)}(\xi)$, the value of ξ depends upon x; in this case, $f^{(n+1)}(\xi)$ must be a polynomial of degree n.)

We next expand the two parts of $R_n(x)$ in terms of Legendre polynomials:

$$\prod_{i=0}^{n} (x - x_i) = a_0 P_0(x) + a_1 P_1(x) + \cdots + a_{n+1} P_{n+1}(x)$$

and

$$\frac{f^{(n+1)}(\xi)}{(n+1)!} = b_0 P_0(x) + b_1 P_1(x) + \cdots + b_n P_n(x)$$

or

$$R_n(x) = \sum_{i=0}^{n+1} a_i P_i(x) \sum_{j=0}^{n} b_j P_j(x) = \sum_{i=0}^{n+1} \sum_{j=0}^{n} a_i b_j P_i(x) P_j(x)$$

We wish $\int_{-1}^{1} R_n(x)\, dx$ to vanish. Part of the vanishing results from the orthogonality of the Legendre polynomials, where

$$\int_{-1}^{1} P_i(x) P_j(x)\, dx = 0, \qquad i \neq j$$

Thus the integral of $R_n(x)$ simplifies to

$$\int_{-1}^{1} R_n(x)\, dx = \sum_{i=0}^{n} a_i b_i \int_{-1}^{1} [P_i(x)]^2\, dx$$

These integrals do not vanish, so we will be forced to make their coefficients vanish. Let us set $a_0 = \cdots = a_n = 0$, leaving only $a_{n+1} \neq 0$. That can be done by requiring that

$$\prod_{i=0}^{n} (x - x_i) = a_{n+1} P_{n+1}(x)$$

which requires that the x_i be the zeros of $P_{n+1}(x)$.

What have we accomplished with all of this? We have shown that with an nth-degree interpolating polynomial, it is possible to get exact results for the integration of all polynomials up through degree $2n + 1$. We have found that the weight factors are given by

$$w_i = \int_{-1}^{1} L_i(x)\, dx$$

and that the abscissas (the x_i) are the zeros of the Legendre polynomial of degree equal to the number of points to be used.

Example 7.5

A 3-point formula ($n = 2$) will give exact results for integrating any polynomial up to and including degree 5. The three points are found by solving

$$P_3(x) = \frac{5x^3 - 3x}{2} = 0$$

which gives $x = 0$ and ± 0.7745966692. The weight factors are found by evaluating

$$\int_{-1}^{1} \prod_{\substack{j=0 \\ j \neq i}}^{2} \frac{x - x_j}{x_i - x_j} dx$$

to find $w_0 = 0.5555555556$, $w_i = 0.8888888889$, and $w_2 = 0.5555555556$. You should verify that the results are correct for integrations of 1, x, x^2, x^3, x^4, and x^5. □

Table 7.3 gives the Gauss-Legendre weights and abscissas through $n = 19$ (20 points). More extensive tables can be found in Abramowitz and Stegun [1964].

Example 7.6

Evaluate $I = \int_{1}^{3} x^x dx$.

First it is necessary to make the substitution $x = z + 2$ in order to get the integral into standard form:

$$I = \int_{-1}^{1} (z + 2)^{z+2} dz$$

The results of using Gauss-Legendre quadrature on this problem are as follows.

N	Points	Estimate of I
1	2	13.125831
2	3	13.696443
3	4	13.724277
4	5	13.725089
5	6	13.725105
9	10	13.725105

In contrast, the trapezoidal rule requires 8193 function evaluations and Simpson's rule requires 129 function evaluations to reach the same accuracy! □

Table 7.3   Abscissas and Weights for Gauss-Legendre Quadrature

Abscissas	N	Weights
	2	
±0.57735 02692		1.0
	3	
0.0		0.88888 88889
±0.77459 66692		0.55555 55556

Table 7.3 (*cont.*)

Abscissas	N	Weights
	4	
± 0.33998 10436		0.65214 51548
± 0.86113 63116		0.34785 48451
	5	
0.0		0.56888 88889
± 0.53846 93101		0.47862 86705
± 0.90617 98459		0.23692 68851
	6	
± 0.23861 91861		0.46791 39346
± 0.66120 93865		0.36076 15730
± 0.93246 95142		0.17132 44924
	8	
± 0.18343 46425		0.36268 37834
± 0.52553 24099		0.31370 66459
± 0.79666 64774		0.22238 10345
± 0.96028 98565		0.10122 85363
	10	
± 0.14887 43390		0.29552 42247
± 0.43339 53941		0.26926 67193
± 0.67940 95683		0.21908 63625
± 0.86506 33667		0.14945 13492
± 0.97390 65285		0.06667 13443
	16	
± 0.09501 25098		0.18945 06105
± 0.28160 35508		0.18260 34150
± 0.45801 67777		0.16915 65194
± 0.61787 62444		0.14959 59888
± 0.75540 44084		0.12462 89713
± 0.86563 12024		0.09515 85117
± 0.94457 50231		0.06225 35239
± 0.98940 09350		0.02715 24594
	20	
± 0.07652 65211		0.15275 33871
± 0.22778 58511		0.14917 29865
± 0.37370 60887		0.14209 61093
± 0.51086 70020		0.13168 86384
± 0.63605 36807		0.11819 45320
± 0.74633 19065		0.10193 01198
± 0.83911 69718		0.08327 67416
± 0.91223 44283		0.06267 20483
± 0.96397 19273		0.04060 14298
± 0.99312 85992		0.01761 40071

7.7.2　Gauss-Chebyshev Quadrature

Another Gauss-type integration scheme is known as *Gauss-Chebyshev quadrature* because it makes use of Chebyshev polynomials in its development. These polynomials are also orthogonal on the interval $[-1, 1]$, but this time with respect to the "weight function" $(1 - x^2)^{-1/2}$. In other words, the more general product

$$\int_{-1}^{1} \frac{T_i(x)T_j(x)}{\sqrt{1-x^2}}\,dx$$

in which $T_i(x)$ designates the ith-degree Chebyshev polynomial, vanishes when $i \neq j$. (The T comes from the older spelling *Tchebycheff*, or sometimes *Tschebyscheff*.) The first few members of this polynomial family are

$$T_0(x) = 1$$
$$T_1(x) = x$$
$$T_2(x) = 2x^2 - 1$$
$$T_3(x) = 4x^3 - 3x$$
$$T_4(x) = 8x^4 - 8x^2 + 1$$

One definition for $T_i(x)$ is

$$T_i(x) = \cos(i\,\mathrm{Cos}^{-1}x)$$

A recurrence formula for their generation, given the first two, is

$$T_i(x) = 2xT_{i-1}(x) - T_{i-2}(x)$$

We will not derive the formulas for Gauss-Chebyshev quadrature. The derivation is similar to that for the Gauss-Legendre method. The weights are given by $\pi/(n + 1)$, where n is the degree of the interpolating polynomial and $n + 1$ is the number of points (note that all weights are equal). The abscissas are given by

$$x_i = \cos\frac{(2i + 1)\pi}{2n + 2}, \qquad i = 0, 1, \ldots, n$$

Table 7.4 contains a short list of these values.

Example 7.7

Evaluate $I = \displaystyle\int_{-1}^{1} \frac{\cos\dfrac{\pi}{2}x\,dx}{\sqrt{1 - x^2}}$ using three points.

$$I = 0.78539\,81634\left\{\cos[\pi(-0.70710\,67812)/2] + \cos(0) + \cos\left[\frac{\pi(0.70710\,67812)}{2}\right]\right\}$$

$$= 1.48285\,6614$$

This compares favorably with the true value of 1.4828355519.　□

Table 7.4 Abscissas and Weights for Gauss-Chebyshev Quadrature

Abscissas	N	Weights
	1	
$\pm 0.70710\ 67812$		$\dfrac{\pi}{2}$
	2	
$\pm 0.86602\ 54038$		$\dfrac{\pi}{3}$
0.0		$\dfrac{\pi}{3}$
	3	
$\pm 0.92387\ 95325$		$\dfrac{\pi}{4}$
$\pm 0.38268\ 34324$		$\dfrac{\pi}{4}$
	4	
$\pm 0.95105\ 55163$		$\dfrac{\pi}{5}$
$\pm 0.58778\ 52523$		$\dfrac{\pi}{5}$
0.0		$\dfrac{\pi}{5}$
	5	
$\pm 0.96592\ 58263$		$\dfrac{\pi}{6}$
$\pm 0.70710\ 67812$		$\dfrac{\pi}{6}$
$\pm 0.25881\ 90451$		$\dfrac{\pi}{6}$
	6	
$\pm 0.97492\ 79122$		$\dfrac{\pi}{7}$
$\pm 0.78183\ 14825$		$\dfrac{\pi}{7}$
$\pm 0.43388\ 37391$		$\dfrac{\pi}{7}$
0.0		$\dfrac{\pi}{7}$

7.7.3 Gauss-Laguerre Quadrature

A third Gauss method applies to integrals of the form

$$I = \int_0^\infty e^{-x} f(x)\, dx = \sum_{i=0}^{m} w_i f(x_i)$$

This is the first method we have seen that applies to an improper integral having an unbounded interval of integration. Although improper integrals with an unbounded upper limit can be transformed into ones over a bounded region, it may be more convenient to use a method such as Gauss-Laguerre quadrature. The derivation of the method will not be given, but it is similar in flavor to the one for Gauss-Legendre quadrature. In this case the functions are the Laguerre polynomials, which are orthogonal with respect to the weight function e^{-x} over the unbounded interval $[0, \infty)$. Table 7.5 contains a list of weights and abscissas up through 15 points.

Example 7.8

Evaluate $I = \displaystyle\int_0^\infty e^{-x} \sin x \, dx$ using 6 points.

$$I = \sum_{i=0}^{5} w_i \sin x_i = (0.45896\,46740)(\sin 0.22284\,66042) + \cdots$$

$$= 0.50004\,94748$$

This is remarkably good for only 6 points between zero and infinity. The exact value is $\frac{1}{2}$, which we can verify by evaluating the known antiderivative. The 15-point Gauss-Laguerre quadrature gives a value of $0.50000\,000021$ rounded to 11 digits.   □

If we can evaluate the antiderivative, then there is little need for numerical integration, but most often we have no antiderivative. In that case we usually have no idea of the accuracy of the answer unless we repeat the numerical integration process several times with increasing numbers of points in an attempt to determine if the process is converging. If we have a rough idea of the accuracy, we can often provide a very accurate result by means of the following procedure: Break the integral into two parts at such a value of x that e^{-x} is close to the round-off error and then make a substitution of $y = x - x_0$ in the second integral, where x_0 is the point at which the break is made. The finite-interval integrand can be evaluated by one of the methods for which an accurate error estimate can be made (e.g., SIMPSON or the adaptive program QUAD). In the case of this example, we shall make the break at 6π, for which $e^{-x} = 6.5 \times 10^{-9}$. Upon making the substitution for x in the infinite-interval integral, we find that

$$\int_{6\pi}^\infty e^{-x}\sin x \, dx = \int_0^\infty e^{-(y+6\pi)}\sin(y + 6\pi)\, dy = e^{-6\pi}\int_0^\infty e^{-y}\sin y \, dy$$

so that the value is $(6.5124121 \times 10^{-9}) \times (0.5000494748) = 3.2365283 \times 10^{-9}$. The value of the integral from 0 to 6π produced by QUAD is 0.4999999967438 (estimated error of 2.7×10^{-13}) so that the sum of the two integrals is in error by about 2.0×10^{-11}. Even if the second integral were in error by 10%, it would not affect the overall accuracy by very much.

Table 7.5 Abscissas and Weights for Gauss-Laguerre Quadrature

Abscissas	N	Weights
	2	
0.58578 64366		0.85355 33906
3.41421 35624		0.14644 66094
	3	
0.41577 45567		0.71109 30099
2.29428 03603		0.27851 77336
6.28994 50829		0.01038 92565
	4	
0.32254 76896		0.60315 41043
1.74576 11012		0.35741 86924
4.53662 02969		0.03888 79085
9.39507 09123		0.00053 92947
	6	
0.22284 66042		0.45896 46739
1.18893 21017		0.41700 08308
2.99273 63261		0.11337 33821
5.77514 35691		0.01039 91975
9.83746 74184		0.00026 10173
15.9828 73981		0.00000 08985

7.7.4 Gauss-Hermite Quadrature

The final Gauss quadrature method at which we shall look is Gauss-Hermite quadrature, which applies to integrals of the form

$$\int_{-\infty}^{\infty} e^{-x^2} f(x)\, dx = \sum_{i=0}^{m} w_i f(x_i)$$

Here the values of the abscissas are found as the zeros of the Hermite polynomials. The first few members of this family are

$$H_0(x) = 1$$

$$H_1(x) = 2x$$

$$H_2(x) = 4x^2 - 2$$

$$H_3(x) = 8x^3 - 12x$$

$$H_4(x) = 16x^4 - 48x^2 + 12$$

The family of polynomials can be generated by a recurrence formula, given the first two members:

$$H_{n+1}(x) = 2xH_n(x) - 2nH_{n-1}(x)$$

A short table of the values of the abscissas and weights is given in Table 7.6.

Table 7.6   Abscissas and Weights for Gauss-Hermite Quadrature

Abscissas	N	Weights
	1	
±0.70710 67811		0.88622 69255
	2	
±1.22474 48714		0.29540 89752
0.0		1.18163 59006
	3	
±1.65068 01239		0.08131 28354
±0.52464 76233		0.80491 40900
	4	
±2.02018 28705		0.01995 32421
±0.95857 24646		0.39361 93232
0.0		0.94530 87205
	5	
±2.35060 49737		0.00453 00099
±1.33584 90740		0.15706 73203
±0.43607 74119		0.72462 95952
	6	
±2.65196 13568		0.00097 17812
±1.67355 16288		0.05451 35828
±0.81628 78829		0.42560 72526
0.0		0.81026 46176

Example 7.9

Evaluate $I = \int_{-\infty}^{\infty} x^2 e^{-x^2}\, dx$ using two points.

$$I = (-0.7071067811)^2(0.8862269255) + (0.7071067811)^2(0.8862269255)$$
$$= 0.8862269255$$

This is the exact value of the integral to the number of significant digits shown ($\pi^{1/2}/2$). This should be no surprise because x^2 is a polynomial of degree 2 and the 2-point formula ($n = 1$) will give exact results on polynomials up through degree 3.   □

7.7.5   Error Terms

The error terms for all of the Gauss quadrature formulas just shown are given in advanced texts in numerical analysis (e.g., Ralston and Rabinowitz [1978]). The error term for Gauss-Legendre quadrature, for example, is

$$E_n(x) = \frac{2^{n+3}[(n+1)!]^4}{(2n+3)[(2n+2)!]^3} f^{(2n+2)}(\xi), \qquad -1 < \xi < 1$$

For the most part the error terms are not of as great practical use as, for instance, the Newton-Cotes error terms are in constructing self-evaluating error estimates.

7.8   Adaptive Quadrature

Consider a function such as the one whose graph is shown in Figure 7.5. We would like to find the integral of this function on the interval [0, 10] with no more error than EPS. Now if we choose a step size H that will give good accuracy near 10, we will find that we do unnecessary work near 0 because in that region we can get high accuracy with large steps. If we think of efficiency in terms of the number of evaluations required of $f(x)$, the integrand function, then it is not very efficient to use the small step size everywhere that is required near 10.

Suppose we write a program that allows the computer to choose the step size based on its estimate of the local error by some method analogous to that used in SIMPSON. We then can accept the value computed over a subinterval $[x_i, x_{i+1}]$ if the estimated error for the interval is less than the allowable error times the ratio of the length of the subinterval to the length of the interval of integration. If the error for this subinterval is not satisfactory, we can then subdivide the subinterval and try again. This process guarantees that the total error is less than the allowable error, barring unusual behavior of the integrand. This gives us an absolute error criterion, which might or might not be acceptable.

We might prefer a relative error criterion for reasons that are similar to those that we discussed in connection with root finding. An absolute error criterion may be far too tight or far too loose for satisfactory results. Unfortunately, a relative error criterion also causes complications because (1) the value of the integral is not available until the integration is finished, and (2) an integrand may oscillate so that the final result may be near zero although partial results may themselves be large. For example, if we integrate sin x from 0 to 20π + DELTA, where DELTA is very small, we have no idea what the value of the integral is near 0, so that it is difficult to know how to apply an error criterion in the relative sense.

FIGURE 7.5   Graph of $f(x) = 0.5 + 0.5(1 - e^{-0.05x}) \sin x^2$.

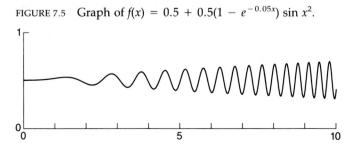

Several other things can go wrong, of course. For well-behaved integrands, the process can work very well, but for any function with a singularity in the domain of integration (or with a singularity in a low-ordered derivative), the attempted subdivision of the interval may never cease. This means that an arbitrary limit must be placed on the level of subdivision in order to prevent bankruptcy. Also, for reasons stated previously, there may be error requirements that cannot possibly be satisfied. Fortunately, most integrands found in practice are well-behaved and can be handled by an automatic integrator, given skill and understanding on the part of the user of the integration programs.

Procedure QUAD is an example of an automatic integrator based on an 8-panel (9-point) closed Newton-Cotes formula. It starts by attempting a 9-point integration over the entire interval of integration and then estimates the error by applying the 8-panel procedure to the left and right half-intervals. The error term for this order of integration contains h^{11}, so that the local error should be less in the 16-panel integration by a factor of 2048. However, since there are now two such local errors, the effective reduction in error is about 1024. Thus we can estimate the error for the 16-panel integration as 1/1023 of the difference between the two estimates. (This is the same idea as the one used in SIMPSON to estimate the error in I2N.)

QUAD uses 8-by-30 arrays to store functional values and abscissas so that nothing has to be computed twice. The 30 is a limitation on the number of levels of subdivision, or *split*, allowed at any point in the program. The program is not aborted if that level is reached; rather, the computation is continued but a count is kept in a variable FLAG of the number of subintervals that did not pass the accuracy test because they could not be further subdivided.

The number of function evaluations permitted is also limited, in this case to 5000. If the number needed will exceed the number permitted, the level of allowed subdivision is reset so that the integration can be continued with degraded accuracy within the permitted number of function evaluations. In such a case, FLAG will also contain a fractional part, which is the fraction of the integration interval remaining to be covered accurately when this occurred.

The call to procedure QUAD is in the following form:

QUAD (A, B, RESULT, ERREST, ABSERR, RELERR, FLAG) ;

A and B are the (real) lower and upper limits of the integration interval. RESULT is a return parameter (real) containing the estimated value of the integral. ERREST (real) is a return parameter containing the estimated error. ABSERR and RELERR (both real) are call parameters specifying the desired absolute and relative errors. FLAG is a (real) return parameter whose integer part contains the number of intervals that were used with less-than-desired accuracy of integration and whose fractional part contains the fractional part of B − A that was not covered at full accuracy.

The error specifications and the error estimate deserve some special mention. ABSERR specifies the desired maximum absolute error, and RELERR specifies the desired maximum relative error. ERREST, which is reported to the user, is the sum of the absolute values of the $(I_{2n} - I_n)/1023$ error estimates at the acceptance level. ABSERR and RELERR are used to establish an error tolerance that this increment of error must pass: the maximum of either ABSERR or RELERR times the absolute value of the current estimate of the integral times the fraction of B – A that the current subinterval represents. If, at the end of the integration, the value of ERREST is below round-off with respect to the absolute value of RESULT, the value of ERREST is doubled and tested again until ABS(RESULT) $\neq$ ABS(RESULT) + ERREST. It would be deceiving to the user to report less error than could occur (and most likely does) through round-off.

It will probably be troublesome to use ABSERR and RELERR much below a small multiple of the machine epsilon times the estimate of the absolute value of the integral. Table 7.7 shows the results of using QUAD on the function in Figure 7.5:

$$f(x) = 0.5 + 0.5\,(1 - e^{-0.05x})\,\sin x^2$$

on the interval [0, 10]. The program was run on a machine using 64-bit floating-point words. The values of ABSERR and RELERR were both set to 1.0×10^{-n}, where $n = 12, 13, \ldots, 19$. Note that at each value of n where the requested error was above the machine epsilon, the reported error (ERREST) was satisfactory. At $n = 16$ and at $n = 17$, the round-off error adjustment was made. At $n = 18$, the error conditions could no longer be met, and FLAG shows that there were three subintervals that did not meet specifications. QUAD got 98.4% of the way across the interval before the limitation on number of function evaluations was imposed. At $n = 19$, FLAG reports that 202 intervals had not met the error criteria, and disaster struck at only 0.8% of the way across the interval. Of course, this was a ridiculous condition to impose, given the size of the computer word.

The last example of the use of QUAD is on an integral with a singularity in the interval of integration. The integrand function is

$$f(x) = \frac{1}{\sqrt{|x - \pi|}}$$

Table 7.7 Results of QUAD for ABSERR = RELERR = 1.0×10^{-n} for Various Values of n

n	RESULT	ERREST	NO. FUN.	FLAG
12	5.00382 76016 7324	4.8943 E-013	1009	0.00000
13	5.00382 76016 7322	4.5358 E-014	1233	0.00000
14	5.00382 76016 7322	5.7508 E-015	1473	0.00000
15	5.00382 76016 7323	6.5323 E-016	1921	0.00000
16	5.00382 76016 7323	5.5276 E-016	2369	0.00000
17	5.00382 76016 7322	4.9233 E-016	2897	0.00000
18	5.00382 76016 7322	7.3681 E-016	3809	3.01563
19	5.00378 43462 1211	4.3861 E-005	4865	202.99105

The integral from 0 to 5 exists (see Exercise 8), but it cannot be evaluated numerically in this form. QUAD gives the following results (using 64-bit arithmetic):

```
RESULT  =  6.52222700047278 E+000
ERREST  =  2.91119699893698 E-003
NUMBER OF FUNCTION EVALUATIONS = 4505
FLAG = 231.37189
```

The fraction in FLAG shows that the integration still had 37% of the way to go; the evaluation of "good" intervals stopped at $x = 3.14055$. (If you do not know about the singularities in the integrand, FLAG will help you find their approximate locations.)

7.9 Dealing with Singularities

Several kinds of misbehavior of the integrand function $f(x)$ create troubles in numerical integration. The presence of a point at which $|f(x)| \to \infty$ or a point at which $f(x)$ is undefined will generally cause the numerical integration to blow up (usually because of division by zero or creating a number out of range). In some cases the integral does not exist and there is nothing to be done. For example,

$$I = \int_0^1 x^{-2}\,dx$$

does not exist because $\lim_{\epsilon \to 0^+} \int_\epsilon^1 x^{-2}\,dx$ does not exist. Nonetheless,

$$I = \int_0^1 x^{-1/2}\,dx$$

does exist because

$$\lim_{\epsilon \to 0^+} \int_\epsilon^1 \frac{dx}{\sqrt{x}} = \lim_{\epsilon \to 0^+} \left(2 - 2\sqrt{\epsilon} \right) = 2$$

(We define the value of the integral in such a case as the value of the limit if the limit exists.)

In other cases the integrand may contain what is sometimes called a *removable singularity*. For example,

$$\int_{-1}^1 \frac{\sin x}{x}\,dx$$

cannot be integrated numerically (if an integration point happens to be precisely zero, we will get a machine halt). However, the limit

$$\lim_{\epsilon \to 0^+} \left\{ \int_{-1}^{-\epsilon} \frac{\sin x}{x}\,dx + \int_\epsilon^1 \frac{\sin x}{x}\,dx \right\}$$

does exist. Because lim sin $x/x = 1$, we find it convenient to define a new function:

$$f(x) = \begin{cases} \dfrac{\sin x}{x}, & x \neq 0 \\ 1, & x = 0 \end{cases}$$

In PASCAL this can be done as

```
FUNCTION SINC (X: REAL): REAL;

BEGIN
   IF X = 0 THEN SINC := 1.0
   ELSE SINC := SIN(X)/X
END;
```

Electrical engineers are familiar with the sinc function as a function that occurs in communication theory.

Where the integral does exist, a singularity of the type found in $\int_0^1 x^{-1/2}\,dx$ can be handled by a change of variable. If, for

$$\int_0^b \frac{f(x)}{x^a}\,dx, \quad a < 1 \quad \text{and} \quad |f(0)| = M < \infty$$

then the integral will exist and a transformation of the form $x = y^{1/(1-a)}$ will reduce $f(x)x^{-a}\,dx$ to

$$f(y^{1/(1-a)})\frac{dy}{1-a}$$

and the integral becomes

$$\int_0^{b^{(1-a)}} f(y^{1/(1-a)})\frac{dy}{1-a}$$

Provided that this causes no difficulties with the transformed f, the numerical integration can be performed.

Example 7.10

Find $I = \displaystyle\int_0^{\pi/2} \frac{\cos x}{\sqrt{x}}\,dx$.

$a = \frac{1}{2}$, so $x = y^2$, and $dx = 2y\,dy$. Thus

$$I = 2\int_0^{\sqrt{\pi/2}} \cos y^2\,dy \quad \square$$

These are only a few examples of the kinds of things that can be done with "bad" integrals. A source of other ideas for handling such kinds of problems is Acton [1970].

7.10   IMSL Function DCADRE

The FORTRAN function DCADRE approximates the integral of a function over a finite interval. The method is one in which Romberg extrapolation is used over the interval or over subintervals if necessary. DCADRE often will behave well and give good results on integrands with simple jump discontinuities.

   The call to DCADRE is the usual FORTRAN function call in which the value of the integral is associated with the name DCADRE.

   DCADRE (F, A, B, AERR, RERR, ERROR, IER)

F is the integrand function, which must have a single argument, must be defined in the calling program (or be a FORTRAN intrinsic function), and must be declared EXTERNAL in the calling program. (The last requirement is necessary if a function is to be used as an argument in a subprogram call.) A and B are the endpoints of the interval of integration. DCADRE attempts to satisfy the condition that the absolute value of the error in computing the integral is less than the maximum of (1) AERR and (2) RERR times the estimated value of the integral. AERR must be nonnegative or it may be zero. RERR must be in the interval [0.0, 0.1]. The IMSL manual suggests that RERR must be large enough that

   100.0 + RERR > 100.0

which means that RERR should be at least 100 times as large as the machine epsilon. The parameter ERROR returns the estimated absolute error in the computation of the integral.

   The error parameter IER returns terminal errors or warnings:

IER =   0   Normal termination.
IER =   65   (Warning) One or more singularities were successfully handled.
IER =   66   (Warning) One or more subinterval results were accepted because the error was small, even though no regular behavior was recognized.
IER = 131   Failure for lack of internal working storage.
IER = 132   Failure because of too much noise in the function relative to error requirements.
IER = 133   RERR > 0.1, RERR < 0.0, or RERR too small for machine precision.

   The original author [de Boor, 1971] remarked that a very cautious person would accept the results if IER were 0 or 65. The merely reasonable person would accept the results if IER = 66, but the truly adventurous person might often be right in accepting the results if IER were 131 or 132.

   As usual, one should be cautious in case F contains a very-high-frequency periodic component with B − A equal to an integer multiple of the period.

Exercises

1. Derive the trapezoidal rule with the aid of a geometrical construction. (This does not permit the development of error terms, but it is helpful in understanding the nature of the rules. Perhaps you can be ingenious in developing the error terms by this method also.)

2. Derive Simpson's $\frac{1}{3}$ rule with the aid of a geometrical construction. (See Exercise 1 for comments.)

3. Determine the trapezoidal rule formula by the method of undetermined coefficients; that is, write

$$\int_{x_0}^{x_1} f(x)\, dx = af(x_0) + bf(x_1) + E$$

where a and b are coefficients to be determined and E is the truncation error. Require that $E = 0$ for $f(x) = 1$ and for $f(x) = x$. (Note that integration is a linear process, so that if we satisfy these conditions we can satisfy $E = 0$ for a general first-degree polynomial.) Replace $x_1 - x_0$ by h in the formula.

To determine E we can let

$$f(x) = c_0 + c_1(x - x_0) + \tfrac{1}{2}f''(\xi)\,(x - x_0)^2$$

where ξ is the usual (unknown) value of x in the Taylor formula. In the resulting formula we can ignore the c_0 and c_1 terms as well as $af(x_0) + bf(x_1)$ because these must of necessity cancel out.

4. Derive the Simpson's $\frac{1}{3}$ rule formula, using the method of undetermined coefficients; that is, let

$$\int_{x_0}^{x_2} f(x)\, dx = af(x_0) + bf(x_1) + cf(x_2) + E$$

where a, b, and c are coefficients to be determined and E is the truncation error. Require $E = 0$ for $f(x) = 1$, $f(x) = x$, and $f(x) = x^2$. To obtain the final formula, let

$$h = x_2 - x_1 = x_1 - x_0$$

We do not need to impose this condition, but it gives a simpler formula. The formula would still give correct results for a parabola unless two of the points were coincident.

To determine E we can let

$$f(x) = c_0 + c_1(x - x_1) + c_2(x - x_1)^2 + c_3(x - x_1)^3 + \left(\frac{1}{4!}\right) f^{(4)}(\xi)(x - x_1)^4$$

Because of the properties of the Newton-Cotes formulas, we need to include the cubic term in this case. As in Exercise 3, we can equate the c_0, c_1, and c_2 terms to the a, b, and c terms before we go very far in the derivation.

5. Show that the Gauss-Legendre 1-point formula is the same as the rectangle (or midpoint) rule.

6. Find the weights and base points for the 3-point Gauss-Legendre method, using the

method of solving the nonlinear equations as was done in the text for the 2-point case.

7. Show that the column of values $I_{m,2}$ of the Romberg integration tableau are the same as those produced by Simpson's first rule.

8. Evaluate the integral $\int_0^5 \dfrac{dx}{\sqrt{|x - \pi|}}$ used in the example on pages 217–18 to illustrate the behavior of QUAD for an integral with a singularity. Note that if we make the substitution $|x - \pi| = y^2$, then we must be careful to use $y^2 = x - \pi$ when $x > \pi$ and $y^2 = \pi - x$ when $x < \pi$.

9. Derive the Richardson extrapolation formulas for Simpson's first rule, $E_{2m} \simeq (I_{2m} - I_m)/15$ and $I^* \simeq (16I_{2m} - I_m)/15$.

10. Integrate the following functions over the indicated intervals, with an error of no more than 1.0×10^{-10}. Choose at least two methods and compare results.

    (a) $\sin x/x$, $[0, 2]$
    (b) $e^{-x}/x^{1/2}$, $[0, 5]$
    (c) $x^2 \sin x/(1 - x)^{1/2}$, $[0, 1]$

11. Evaluate the Gauss-Chebyshev example, using 5 points.

12. Repeat Exercise 11, using Gauss-Legendre integration with 5 points. Compare the result with that of Exercise 11.

13. The gamma function $\Gamma(n)$ is defined by the integral

$$\Gamma(n) = \int_0^\infty x^{n-1}e^{-x}\,dx, \qquad 0 < n$$

If n is a positive integer, $\Gamma(n + 1) = n!$. $\Gamma(1)$ is defined as 1. Also, it is possible to show that $\Gamma(n + 1) = n\Gamma(n)$, whether n is an integer or not. For these reasons we need only a table of $\Gamma(n)$ for $1 \leq n \leq 2$ for practical purposes.
    Produce a table of $\Gamma(n)$ for $n = 1.0, 1.1, 1.2, \ldots, 2.0$.

14. (a) Transform the integral $\int_0^\pi \sin x\,dx$ to the interval of integration $[-1, 1]$ by means of a linear transformation and evaluate using Gauss-Legendre quadrature. How many points are necessary to obtain 10 decimal digits of accuracy?

    (b) Evaluate the same integral by multiplying and dividing the transformed integrand by $\sqrt{1 - x^2}$ so that the integral can be evaluated using Gauss-Chebyshev quadrature. How do the results compare with those from part (a)? Why does this not work as well? (Think about the nature of the integrands that can be evaluated exactly using Gaussian quadrature.)

15. The spectral distribution of the energy flux from a black body radiating electromagnetic energy is given by

$$\frac{a\lambda^{-5}}{e^{b/\lambda T} - 1}$$

where $a = 3.74 \times 10^{-5}$ erg·cm²/s and $b = 1.4387$ cm·K, λ is the wavelength of the radiation in centimeters, and T is the temperature of the black body in degrees Kelvin.

    (a) The radiation from the sun can be approximated as black-body radiation with its peak energy flux at $\lambda = 6000$ Å, where 1 Å $= 10^{-8}$ cm. Find the equivalent black-body temperature of the sun.

(b) Find the relative amounts of energy emitted by the sun in the ultraviolet ($\lambda <$ 4000 Å), in the infrared ($\lambda > 8000$ Å), and in the visible (4000 Å $< \lambda <$ 8000 Å) regions of the solar spectrum. (*Hint:* Replace wavelength λ by frequency f by the use of the relation $\lambda f = c$, where c is the speed of light in a vacuum. The factor $1/(e^z - 1)$, where $bf/cT = z$, in the integrand can be transformed into $e^{-z}/(1 - e^{-z})$, which will allow evaluation of the integral over the entire spectrum to be done by Gauss-Laguerre integration. To evaluate the integral over the ultraviolet portion of the spectrum, make a translation of the origin to the beginning of the ultraviolet region: $f' = f - f''$, where f'' is the frequency at which the ultraviolet begins, f' is the new frequency variable, and f is the old frequency variable in z. Then use Gauss-Laguerre integration once more.)

16. An exponential horn is a type of acoustic transformer that couples a loudspeaker (driver) efficiently to the air. The cross-sectional area of the horn increases exponentially with the distance from the driver. In order to estimate the amount of material required to make the horn, it is necessary to calculate the surface area of the horn and multiply that by the thickness.

    A certain horn couples a driver with a 10-inch diameter to the mouth, which is 4 feet in diameter. The length of the horn along the axis from the throat (where the driver is) to the mouth is 8 feet. Find the inside surface area of the horn.

17. A very important function in communication theory is the sinc function, defined one way as

$$\text{sinc}(x) = \begin{cases} \dfrac{\sin ax}{x}, & -\infty < x < 0, \quad 0 < x < \infty \\ a & x = 0 \end{cases}$$

    It can be shown that the energy density versus frequency of a rectangular pulse signal is given by

$$\text{sinc}^2(\omega)$$

    where ω is the angular frequency $2\pi f$.

    Find the relative amounts of energy contained in the main frequency lobe ($-\pi/a < \omega < \pi/a$), the first side lobes ($-2\pi/a < \omega < -\pi/a$ and $\pi/a < \omega < 2\pi/a$), and the second side lobes.

18. The Gauss error function $\text{erf}(x)$ is

$$\text{erf}(x) = 2/(\pi^{1/2}) \int_0^x e^{-t^2} dt$$

    The function of $\text{erf}(x)$ is closely related to the normal probability function with mean zero and unit variance.

$$P(x) = \frac{1}{\sqrt{2\pi}} \int_{-\infty}^x e^{-t^2/2} dt$$

    It also appears in other places, such as in the solution of heat conduction problems in unbounded media in rectangular coordinates.

    Compute a table of values of $\text{erf}(x)$ for $0 \leqslant x \leqslant 10$ by units of 0.1. Compare these numbers (say, at unit intervals) with the values of the normal probability function (one set or the other must be transformed). Note that $P(0) = 0.5$.

19. When light or other electromagnetic waves pass through a slit the diffraction pattern at a small distance (a small fraction of the slit width) depends upon two integrals known as the Fresnel integrals:

$$C = \int_0^v \cos\left(\frac{\pi u^2}{2}\right) du \qquad S = \int_0^v \sin\left(\frac{\pi u^2}{2}\right) du$$

   (a) Produce graphs of C and S for v from 0 to 5.
   (b) Plot the values of C as the abscissa and S as the ordinate with v as a parameter to produce what is known in optics as Cornu's spiral.

20. Evaluate numerically the integral

$$\int_0^\infty \frac{dx}{(x^2 + 1)^{3/2}}$$

   and estimate the accuracy of your result. *Suggestion:*
   (a) Use QUAD to evaluate

$$\int_0^{10} f(x)\, dx$$

   and use the error estimate from QUAD for this part of the integral.
   (b) Estimate

$$\int_{10}^\infty f(x)\, dx$$

   by approximating $f(x) = (x^2 + 1)^{-3/2} \simeq x^{-3}$. The relative error in this approximation is no worse than the worst relative error in approximating $f(x)$ (about 1.5%, which you can calculate at $x = 10$).
   (c) The absolute errors of the two parts can now be added. The absolute error in part (b) is approximately the relative error times the estimated value of the integral evaluated in (b).

Problems

PROJECT PROBLEM 7.1

1. Estimate the value of π by numerically integrating to find the first-quadrant area inside the circle $x^2 + y^2 = 1$. The area has the value $\pi/4$.
      Why does the value not seem to converge on π?
2. Replace the problem in (1) by the following problem shown in Figure 7.6. Approximate π by finding 8 times the area A of the sector with angle 45°. Use the same integrand as in (1), but integrate only from 0 to $2^{-1/2}$ and then subtract the area of triangle B to find the area A.
      Why does this work better?

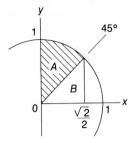

FIGURE 7.6   Geometry necessary for determining an
approximation to π by numerical integration.

PROJECT PROBLEM 7.2

Write a program to provide Romberg integration controlled from the keyboard
so that it first automatically produces a 5×5 triangular array of values of the
$I_{m,n}$. If convergence is not as yet satisfactory, allow for intervention from the
keyboard to produce one or more additional rows of the array as desired.

Test the software on integrands that can be evaluated by classical methods.

PROJECT PROBLEM 7.3

Write an interactive program for integration using QUAD. The program should
question the user concerning the endpoints of the integration interval and the
accuracy desired. The results reported should include the estimate of the value
of the integral, the estimated error, the number of function evaluations, and the
value of FLAG.

Depending on how your PASCAL system works, you will need an already-
compiled function FUN or you will need a source code function FUN whose source
file name is contained in an *include* statement.

PROJECT PROBLEM 7.4

Figure 7.7 shows a simple RC electrical circuit known as a two-port circuit because
it has four terminals instead of two. It can be shown [Nilsson, 1985] that if the
input port is driven by a voltage impulse function $f(t) = \delta(t)$, which can be
described as

$$\delta(t) = 0, \qquad t \neq 0$$

$$\int_{-\infty}^{\infty} \delta(t)\, dt = 1$$

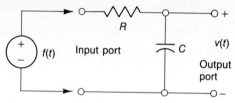

FIGURE 7.7   Two-port network.

that the response (output) voltage $v(t)$ is

$$v(t) = h(t) = e^{-t/RC}u(t)$$

Where R is the resistance in ohms, C is the capacitance in farads, and t is the time in seconds. $u(t)$ is called the *unit step function*

$$u(t) = \begin{cases} 1, & t \geq 0 \\ 0, & t < 0 \end{cases}$$

It can also be shown [Gabel and Roberts, 1980] that the output for any other input signal $f(t)$ is given by the convolution integral

$$v(t) = \int_{-\infty}^{\infty} f(\tau)h(t - \tau)\,d\tau = \int_{-\infty}^{\infty} f(t - \tau)h(\tau)\,d\tau$$

where $h(t)$ is the impulse response function for the circuit and $f(t)$ is the driving function.

Find the response (output voltage $v(t)$) for this circuit for the following driving functions:

1. $f(t) = 1$, $0 < t < 1$; $f(t) = 0$ otherwise.
2. $f(t) = t(1 - t)$, $0 < t < 1$; $f(t) = 0$ otherwise.
3. $f(t) = \sin(\pi t)$, $0 < t < 1$; $f(t) = 0$ otherwise.

Graph both the input function $f(t)$ and the output function $v(t)$ on the same axes. Use $R = 2\ \Omega$ and $C = 1$ F. It will be necessary to evaluate an integral for each point on the graph of $v(t)$, so you may wish to experiment with using Gauss integration for speed. Note that Gauss-Laguerre integration is not necessary (or desirable) because the domain over which $f(t)$ is nonzero is bounded. It is best not to incorporate $u(t)$ into $h(t)$; rather, it is desirable to let the limits of integration contain the parameter t.

Comment: It is possible to produce the same results using classical mathematics, but for many $f(t)$ the integration is difficult if not impossible.

PROJECT PROBLEM 7.5

In Project Problem 5.2 we found an expression for the normalized volume x (= V/V_0) of a simplified model of a natural gas reservoir with water drive and gas entrapment as a function of the normalized pressure y ($= P/P_0$) with parameters F, b, and k. It can be shown [Hultquist and Monash, 1979] that the ultimate recovery u (the fraction of the initial gas in place recovered during the reservoir life) is given by

$$u = \frac{1 - x_c y_c}{1 + F} + \frac{F}{1 + F} \int_{y_c}^{1} x\, dy$$

where the expression for x is that given in Project Problem 5.2 and where x_c and y_c are the coordinates of the cutoff point found in that problem.

Find the ultimate recovery of a gas reservoir for the various values (in all combinations) of F, C, r_0, x_{co}, and y_{co} found in Project Problem 5.2.

PROJECT PROBLEM 7.6

The gamma probability distribution has a probability density function $f(x)$ given by

$$f(x) = \begin{cases} \dfrac{\beta^{-\alpha} x^{\alpha-1} e^{-x/\beta}}{\Gamma(\alpha)}, & \text{if } x > 0 \\ 0, & \text{otherwise} \end{cases}$$

where

$$\Gamma(\alpha) = \int_{0}^{\infty} t^{\alpha-1} e^{-t}\, dt$$

In order to evaluate α and β for a set of data $\{x_i\}$, we can use the maximum-likelihood estimator method, which leads to the system of simultaneous equations

$$\alpha\beta = \bar{x}$$

$$\ln \beta + \varphi(\alpha) = \left(\frac{1}{n}\right) \sum_{i=1}^{n} \ln (x_i)$$

where $\varphi(\alpha)$ is the derivative of $\Gamma(\alpha)$ with respect to α and $\bar{x}$ is the mean of the set of data:

$$\bar{x} = \left(\frac{1}{n}\right) \sum_{i=1}^{n} x_i$$

One expression for the function $\varphi(z)$ is

$$\varphi(z) = \ln z - \frac{1}{(2z)} - 2 \int_{0}^{\infty} \frac{t\, dt}{(t^2 + z^2)(e^{2\pi t} - 1)}$$

provided that the real part of z is positive (which is true here).

1. Multiply the numerator and denominator of the integrand in the expression for $\varphi(z)$ by $e^{-2\pi t}$ and then let $s = 2\pi t$ in order to put the integral into a form suitable for Gauss-Laguerre quadrature.

2. Write a PASCAL function PSI (Z) to evaluate $\varphi(z)$ by means of a 15-point Gauss-Laguerre formula. The points and weights can be found in Abramowitz and Stegun [1964], Chapter 25. The values of $\varphi(z)$ that you compute can be checked against the table found in Chapter 6 of Abramowitz and Stegun ($\varphi(z)$ is also known as the *digamma function*).

3. Create a PASCAL function PSIP (Z) for the derivative of $\varphi(z)$. You can derive an expression for this function by differentiating the expression for $\varphi(z)$ with respect to z. This will require that you differentiate the integral with respect to the parameter z. (See any advanced calculus text for a discussion of differentiation under the integral sign.) This function can be evaluated with only a low-order (5 or 6 points) Gauss-Laguerre formula because Newton's method does not require high accuracy in the derivative term.

4. Substitute $\beta = \bar{x}/\alpha$ from the first equation into the second equation to give an equation that can be solved for α by Newton's method. Use the following set of data:

3.57	3.26	5.48	2.19
1.99	10.56	0.73	7.32
1.15	6.03	2.65	4.37
7.32	3.87	12.76	4.75
3.47	0.05	8.08	3.14

Listings

LISTING OF SIMPSON

```
PROCEDURE  SIMPSON (      A:      REAL;
                         B:      REAL;
                         N:      INTEGER;
             VAR   I1N:      REAL;
             VAR   I2N:      REAL;
             VAR   ISTAR:    REAL;
             VAR   E2N:      REAL      );
```

```
{SIMPSON is a two-stage Simpson's Rule integrator that
produces two estimates of the value of the integral:
an error estimate for the more accurate value, and an
extrapolated estimate that is usually much more accurate
than either of the computed values.

Parameters are:

    A - Lower limit of the definite integral (input)
```

B - Upper limit of the definite integral (input)

N - Number of panels to be used in the more accurate evaluation of the integral (I2N); N must be a multiple of 4   (output)

I2N - Simpson's Rule value using N panels (output)

I1N - Simpson's Rule value using N/2 panels (output)

ISTAR - Extrapolated estimate of the integral using ISTAR = (16*I2N - I1N)/15 (output)

E2N - Error of I2N estimated from E2N = (I2N - I1N)/15 (output)

Note: The integrand function must be supplied by the user in the form of a function named FUN. This function must be included in the calling program explicitly or must be loaded using an "include" statement. }

```
VAR    M, I, J:              INTEGER;
       O, OO:                BOOLEAN;
       X, V, SUM10, SUM21:   REAL;
       SUM22, SUM41, SUM42:  REAL;

BEGIN

IF ODD(N) THEN BEGIN WRITELN('Odd N.   Halt.');
                     HALT; END;

IF ODD(N DIV 2) THEN
  BEGIN
    WRITELN('N must be a multiple of 4.  Halt.'); HALT;
  END;

SUM41 := 0.0;
SUM42 := 0.0;
SUM21 := 0.0;
SUM22 := 0.0;

SUM10 := FUN(A) + FUN(B);

FOR I := 1 TO N-1 DO
  BEGIN
    X := A + I*(B - A)/N;
    V := FUN(X);
    J := I DIV 2;
    O := ODD(I);  OO := ODD(J);
```

```
            IF O THEN SUM41 := SUM41 + V
              ELSE
                BEGIN
                  SUM21 := SUM21 + V;
                  IF OO THEN SUM42 := SUM42 + V
                    ELSE SUM22 := SUM22 + V;
                END;
        END;


    SUM41 := 4.0*SUM41;
    SUM42 := 4.0*SUM42;
    SUM21 := 2.0*SUM21;
    SUM22 := 2.0*SUM22;


    I2N := (SUM10 + SUM41 + SUM21)*(B - A)/(3*N);
    I1N := (SUM10 + SUM42 + SUM22)*(B - A)/(1.5*N);


    ISTAR := (16.0*I2N - I1N)/15.0;  {Better estimate than I2N.}


    E2N := (I2N - I1N)/15.0;    {Error estimate for 2N step}
                                {Simpson's Rule.}
    END;
```

LISTING OF QUAD

```
    PROCEDURE QUAD      (   A:  REAL;
                           B:  REAL;
                   VAR  RESULT:  REAL;
                       ERREST:  REAL;
                       ABSERR:  REAL;
                       RELERR:  REAL;
                   VAR   NOFUN:  INTEGER;
                   VAR    FLAG:  REAL);
```

```
{PROCEDURE QUAD is an adaptive Newton-Cotes nine-point
 integration scheme.

 PARAMETERS:

 A:      The lower limit of the interval of integration.

 B:      The upper limit of the interval of integration.

 RESULT: The estimated value of the integral.

 ERREST: An estimate of the magnitude of the error.

 NOFUN:  Number of evaluations of FUN(X) made during the
         integration process.
```

ABSERR: An absolute error tolerance (non-negative).

RELERR: A relative error tolerance (non-negative).

FLAG: Reliability indicator. If FLAG is zero, then
           RESULT is probably good to the least restrictive
           error tolerance.   If FLAG is XXX.YYY (non-zero),
           then XXX is the number of which have not
           converged, and 0.YYY is the fraction of the whole
           interval left to do when the limit on NOFUN was
           reached.

FUN(X):   The user must furnish a function FUN(X) for the
           integrand function.   In those systems which allow
           a function as a parameter in a procedure call,
           FUN can be included in the parameter list.
           Otherwise it may be supplied in an include file.}

```
VAR    W0,W1,W2,W3,W4,AREA,X0:          REAL;
       F0,S1,S2,COR11,TEMP:             REAL;
       QPREV,QNOW,QDIFF,QLEFT:          REAL;
       ESTERR,TOLERR:                   REAL;

       QRIGHT:         ARRAY[1..31] OF REAL;
       F,X:            ARRAY[1..16] OF REAL;
       FSAVE,XSAVE:    ARRAY[1..8,1..30] OF REAL;

       SPLMIN,SPLMAX,SPLOUT:            INTEGER;
       NOMAX,NOFIN,SPL,NIM,I,J,JJ:      INTEGER;

       RTF:                             BOOLEAN;


FUNCTION PWR(X: REAL; N: INTEGER): REAL;

VAR      Y,Z:        REAL;
         BIGN:       INTEGER;
            O:       BOOLEAN;

BEGIN
  Y := 1.0;
  Z := X;
  BIGN := N;

  REPEAT
    BEGIN
      O := ODD(BIGN);
      BIGN := BIGN DIV 2;
```

```
            IF O THEN Y := Y*Z;
            Z := Z*Z;
        END
    UNTIL BIGN = 0;

  PWR := Y
END;

FUNCTION AMAX1(X,Y: REAL): REAL;

BEGIN
  IF X > Y THEN AMAX1 := X
  ELSE
    AMAX1 := Y

END;



PROCEDURE LEVEL_DOWN;

{No convergence . . .}
{Locate the next interval}

BEGIN

NIM := 2*NIM;
SPL := SPL + 1;       {One more level of split}

{Store right-hand elements for the future}

FOR I := 1 TO 8 DO
  BEGIN
    FSAVE[I,SPL] := F[I+8];
    XSAVE[I,SPL] := X[I+8]
  END;



{Store left-hand elements for use now}

QPREV := QLEFT;
FOR I := 1 TO 8 DO
  BEGIN
    J := -I;
    F[2*J+18] := F[J+9];
    X[2*J+18] := X[J+9]
  END
END;

PROCEDURE SET_RETURN;
```

```
{Finalization in preparation for return}

BEGIN

RESULT := RESULT + COR11;

{Make sure that ERREST is not less than roundoff}

IF ERREST <> 0.0 THEN
  BEGIN
    TEMP := ABS(RESULT) + ERREST;
    IF TEMP = ABS(RESULT) THEN
      BEGIN
        REPEAT
          BEGIN
            ERREST := 2.0*ERREST;
            TEMP := ABS(RESULT) + ERREST
          END
        UNTIL TEMP <> ABS(RESULT);
      END;
    END;
    RTF := TRUE
END;

PROCEDURE SUM_UP;

{Add contributions into running sums}

BEGIN
RESULT := RESULT + QNOW;
ERREST := ERREST + ESTERR;
COR11 := COR11 + QDIFF/1023.0;

{Find the next interval}

WHILE NIM <> 2*TRUNC(NIM/2) DO
BEGIN
  NIM := TRUNC(NIM/2);
  SPL := SPL - 1;
END;
NIM := NIM + 1;
IF SPL <= 0 THEN SET_RETURN
ELSE
  BEGIN                    {Gather elements for next interval}
    QPREV := QRIGHT[SPL];
    X0 := X[16];
    F0 := F[16];
    FOR I := 1 TO 8 DO
      BEGIN
        F[2*I] := FSAVE[I,SPL];
```

```
            X[2*I]  := XSAVE[I,SPL]
        END;
    END;
END;


PROCEDURE TROUBLE;

{Trouble; number of function values about to exceed limit}

BEGIN

NOFIN := 2*NOFIN;
SPLMAX := SPLOUT;
FLAG := FLAG +(B - X0)/(B - A);
SUM_UP;
END;


PROCEDURE FLAG_INCR;

{Current level is SPLMAX}

BEGIN
FLAG := FLAG + 1.0;
SUM_UP;
END;


PROCEDURE INITIALIZE;

{Initialization for first interval}

BEGIN

SPL := 0;
NIM := 1;
X0 := A;
X[16] := B;
QPREV := 0.0;
F0 := FUN(X0);
S2 := (B - A)/16.0;
X[8] := (X0 + X[16])/2.0;
X[4] := (X0 + X[8])/2.0;
X[12] := (X[8] + X[16])/2.0;
X[2] := (X0 + X[4])/2.0;
X[6] := (X[4] + X[8])/2.0;
X[10] := (X[8] + X[12])/2.0;
X[14] := (X[12] + X[16])/2.0;
FOR J := 1 TO 8 DO
  F[2*J] := FUN(X[2*J]);
NOFUN := 9;

END;
```

```
PROCEDURE MAIN;

{Central calculation. Calculates X1,X3,...,X15,F1,F3,...,
F15,QLEFT,QRIGHT,QNOW,QDIFF, and AREA from X0,X2,...,X16,
and F0,F2,...,F16}

BEGIN

X[1] := (X0 + X[2])/2.0;
F[1] := FUN(X[1]);
FOR JJ := 1 TO 7 DO
  BEGIN
    J := JJ*2 + 1;
    X[J] := (X[J-1] + X[J+1])/2.0;
    F[J] := FUN(X[J])
  END;
NOFUN := NOFUN + 8;
S1 := (X[16] - X0)/16.0;
QLEFT := (W0*(F0 + F[8]) +
         W1*(F[1] + F[7]) +
         W2*(F[2] + F[6]) +
         W3*(F[3] + F[5]) +
         W4*F[4])*S1;

QRIGHT[SPL + 1] := (W0*(F[8] + F[16]) +
                    W1*(F[9] + F[15]) +
                    W2*(F[10] + F[14]) +
                    W3*(F[11] + F[13]) +
                    W4*F[12])*S1;

QNOW := QLEFT + QRIGHT[SPL+1];
QDIFF := QNOW - QPREV;
AREA := AREA + QDIFF;

END;


{Beginning of PROCEDURE QUAD.  QUAD estimates the integral
 of a user-provided function FUN(X) over the interval
 A to B.  It uses a nine-point Newton-Cotes method
 adaptively in an attempt to satisfy the error tolerances
 specified by the user.}

BEGIN

SPLMIN = 1;
SPLMAX = 30;
SPLOUT = 6;
NOMAX = 5000;
NOFIN = NOMAX  - 8*(SPLMAX - SPLOUT
        + TRUNC(PWR(2.0, SPLOUT+1)));
```

```
{If NOFUN reaches NOFIN then we have trouble}

W0 = 3956.0/14175.0;
W1 = 23552.0/14175.0;
W2 = -3712.0/14175.0;
W3 = 41984.0/14175.0;
W4 = -18160.0/14175.0;

{Initialize sums}

FLAG := 0.0;
RESULT := 0.0;
COR11 := 0.0;
AREA := 0.0;
ERREST := 0.0;
NOFUN := 0;

IF A <> B THEN
  BEGIN
    INITIALIZE;
    RTF := FALSE;

    REPEAT
      BEGIN
        MAIN;

{Interval convergence test}

        ESTERR := ABS(QDIFF)/1023.0;
        TOLERR := AMAX1(ABSERR, RELERR*ABS(AREA))*(S1/S2);

        IF SPL < SPLMIN THEN LEVEL_DOWN
        ELSE
          IF SPL >= SPLMAX THEN FLAG_INCR
          ELSE
            IF NOFUN > NOFIN THEN TROUBLE
            ELSE
              IF ESTERR <= TOLERR THEN SUM_UP
              ELSE LEVEL_DOWN;
      END
    UNTIL RTF = TRUE
  END;

END;
```

8

Numerical Solution of Differential Equations

The subject of differential equations is one that has a long history in the applications of mathematics to scientific and engineering problems. Many of the important physical laws of the universe are expressed in differential equation form, such as Newton's second law of motion, which governs the field of dynamics (e.g., rocket flight, planetary motion, and the behavior of vibrating systems). As an example of a different kind of system, we shall formulate a very simplified ecological problem that illustrates a phenomenon observed in nature.

First we define foxes and rabbits as the only animal inhabitants of a closed ecological system. A simplified rabbit is one that eats grass, procreates, and is eaten by foxes. No rabbits die of old age. All foxes eat rabbits, procreate, and die of geriatric diseases. The grass for the rabbits is in constant supply.

Let $f(t)$ be the number of foxes and $r(t)$ be the number of rabbits at any time t. We take f and r as continuous variables (the numbers of each species are relatively large) and ignore the problem of having to have two sexes of each species. The time rate of change of the number of rabbits is

$$\frac{dr(t)}{dt} = -ar(t)f(t) + br(t)$$

The term $br(t)$ is the birth rate of the rabbits. It expresses the simple fact that the more rabbits there are, the greater the number of births per unit time given a constant environment. The constant b is the unit birth rate. The term $-ar(t)f(t)$ is a death-rate term (because of the minus sign) and assumes that the rate at which rabbits are eaten is proportional to the product of the number of foxes and the number of rabbits.

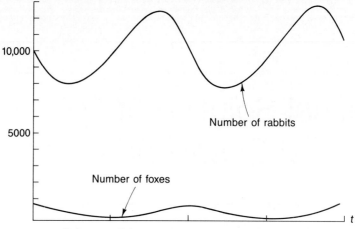

FIGURE 8.1   Behavior of the number of foxes and the
number of rabbits in a simple predator-prey
model.

The rate of change of the number of foxes is given by

$$\frac{df(t)}{dt} = -pf(t) + qf(t)r(t)$$

The term $-pf(t)$ is the death rate of the foxes. The term $qf(t)r(t)$ expresses the
idea that the birth rate of the foxes is dependent not only upon the number of
foxes but also upon the availability of the rabbits as food.

These two differential equations form a system of differential equations that
is very difficult to solve by any means other than numerical solution. An example
of the solution using the constants $a = 2 \times 10^{-6}$, $b = 10^{-3}$, $p = 10^{-2}$, and $q = 10^{-6}$, with $r(0) = 10{,}000$ and $f(0) = 1000$, is shown in Figure 8.1. The solution
shows a cyclic behavior that is often observed in nature in predator-prey systems.

8.1   Differential Equations and Systems

Before we look at numerical methods, we need to define the problems with which
we shall deal. The general case of a single first-order differential equation in
standard form is described by

$$\frac{dy}{dx} = f(x, y), \qquad y(x_0) = y_0$$

The right-hand-side function $f(x, y)$ covers a wide variety of different forms
ranging from $f(x, y) = 0$ to $f(x, y) =$ any imaginable function of x and y. Note
that we do not have a complete problem without the specification of one point

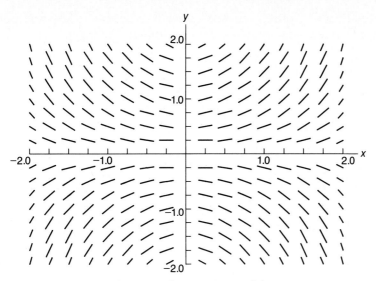

FIGURE 8.2 Tangent lines to solution curves of the differential equation $dy/dx = xy$.

through which the solution must pass. A differential equation by itself specifies a family of solutions, and the condition $y(x_0) = y_0$ picks out one member of that family.

In this book we do not deal with such matters as the proof of existence of a solution for a given differential equation. Proofs of existence are found in texts such as Coddington [1961] and Simmons [1972].

Figure 8.2 shows a region of the xy plane near the origin with a grid of points spaced $\frac{1}{4}$ unit apart. Drawn through each point is a short line segment that represents the tangent line to the solution curve through that point for the differential equation $dy/dx = xy$. It does not take a great deal of imagination to visualize how the solution curves Ce^{x^2} behave throughout the xy plane. One very primitive method of solution, in fact, consists of producing a handmade sketch of the desired curve using the tangent-line information.

At this point we need to introduce terminology that will be useful. A differential equation is an equation containing one or more derivatives or differentials. It is an ordinary differential equation (ODE) if there is only one independent variable (typically the time), and it is a partial differential equation (PDE) if there is more than one independent variable, so that partial derivatives are necessary. We shall deal only with ordinary differential equations in this chapter.

The order of an ODE is the order of the highest-order derivative in the equation. $dy/dx = xy$ is a first-order ODE, but $d^2\theta/dt^2 = (g/L) \sin \theta$ is a second-order equation. An ODE is linear if the dependent variable (y or θ) appears in linear form. The equation may contain no terms involving the dependent variable or its derivatives to any power other than first, and it may contain no products of the dependent variable or its derivatives among themselves, such as $y\, dy/dx$. The first equation is linear because y and its derivatives appear to the first power only. The second equation is nonlinear because $\sin \theta$ is nonlinear:

$$\sin \theta = \theta - \frac{\theta^3}{6} + \frac{\theta^5}{120} + \cdots$$

Other examples of nonlinear equations are

$$y \frac{dy}{dx} = x - y$$

$$\left(\frac{dy}{dx}\right)^2 = y$$

$$\frac{d^2y}{dx^2} + y\frac{dy}{dx} + y^2 = 0$$

Linear equations generally should be solved by other means than numerical integration. There are powerful classical methods for solving linear differential equations. These methods can be found in texts on ordinary differential equations and on circuit analysis.

A system of differential equations consists of two or more equations involving a like number of dependent variables. It is not generally interesting as a system unless there is *cross-coupling*, in which more than one dependent variable occurs in each equation. (Otherwise such a system is really only a set of independent differential equations, which can be solved serially rather than in parallel.) For example, a system might be

$$\frac{dy}{dt} + x = y$$

$$\frac{d^2x}{dt^2} - xy = \sin t$$

A single higher-order ODE and its initial conditions always can be converted to a system of first-order equations with initial conditions. For example,

$$\frac{d^3y}{dx^3} + 18xy\frac{d^2y}{dx^2} + x\frac{dy}{dx} + y^2 = 0$$

$$y(x_0) = y_0, \qquad y'(x_0) = y'_0, \qquad y''(x_0) = y''_0$$

can be converted to a first-order system by means of the substitutions $u = dy/dx$ and $v = du/dx = d^2y/dx^2$. The system then becomes

$$\frac{dv}{dx} + 18xyv + xu + y^2 = 0$$

$$\frac{dy}{dx} = u$$

$$\frac{du}{dx} = v$$

and

$$y(x_0) = y_0$$
$$u(x_0) = y_0'$$
$$v(x_0) = y_0''$$

We should note that the word *order* is used for both equations and systems. An nth-order ODE can be transformed into n first-order equations, in which case the system of equations is called nth order. We should also recall from differential equations that an nth-order initial-value problem requires n initial values. Hence, after transformation the nth-order system has exactly one initial condition for each equation.

An ODE or a system of ODEs with all conditions specified at one value of the independent variable is usually called an *initial-value problem*. This is appropriate terminology if time is the independent variable and all conditions are specified at $t = 0$. In contrast, there are problems in which conditions are specified at other points. For example, a second-order ODE problem can be specified as follows:

$$\frac{d^2y}{dx^2} = f\left(\frac{dy}{dx}, y, x\right)$$
$$y(a) = A, \qquad y(b) = B$$

in which case the problem is called a *boundary-value problem* because the conditions are specified at the boundaries of an interval. Problems of this kind require special treatment, as we shall see later.

8.2 Methods of Solution

One very general method of attacking the solution of an ordinary differential equation is to replace the differential equation

$$\frac{dy}{dx} = f(x, y)$$

by the integral equation

$$\int_{x_i}^{x_{i+1}} dy = y_{i+1} - y_i = \int_{x_i}^{x_{i+1}} f(x, y)\, dy$$

or

$$y_{i+1} = y_i + \int_{x_i}^{x_{i+1}} f(x, y)\, dx$$

where $[x_i, x_{i+1}]$ is an interval over which we wish to integrate. We would start, of course, with the interval $[x_0, x_1]$. Then, if we could find y_1, we would consider (x_1, y_1) as the new starting values in order to continue the process over $[x_1, x_2]$. Our problem is what to do about the y that appears in the integral because we do not know the values of y in the interval of integration. Much of the art of solving differential equations numerically rests on how to approximate $f(x, y)$ in the interval.

The first method at which we will look simply assumes that f is constant over the interval with the value $f(x_i, y_i)$. Then if we let $h = x_{i+1} - x_i$, we have

$$y_{i+1} = y_i + \int_{x_i}^{x_{i+1}} f(x_i, y_i)\, dx = y_i + h f(x_i, y_i)$$

This is known as *Euler's method*. Example 8.1 shows how it works.

Example 8.1

Use Euler's method with $h = 0.1$ to solve the initial-value problem $dy/dx = xy$, $y(0) = 1.0$. (The classical solution is $y = 1.0\, e^{x^2/2}$.)

$f(x, y) = xy$, so that for $x_0 = 0$ and $y_0 = 1$, we have $f(x, y) = 0$. Thus $y_1 = y_0 + 0h = 1.0$. For $x_1 = 0.1$, $y_1 = 1.0$, we have $f_1 = (0.1)(1.0) = 0.1$ and

$$y_2 = y_1 + hf_1 = 1.0 + (0.1)(0.1) = 1.01$$

$$y_3 = y_2 + hf_2 = 1.01 + (0.1)(0.2)(1.01) = 1.0302$$

These values compare with $y(0.1) = 1.005012521$, $y(0.2) = 1.02020134$, and $y(0.3) = 1.046027860$. The errors are rather substantial, but they can be reduced by reducing h, as we shall see. Figure 8.3 shows the Euler solution and the analytical solution.   □

Euler's method can also be derived by using a linear extrapolation of the solution curve. The slope of the extrapolating line is the value of $f(x, y)$ at each point. This is left as an exercise. The geometrical interpretation is important because it helps us understand the sources of the error in a numerical solution.

FIGURE 8.3

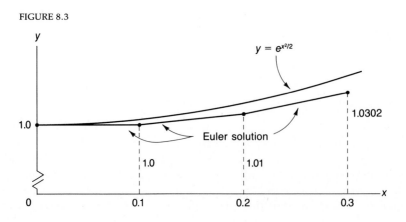

To see how the error in Euler's method behaves, we can derive an error term by expanding the integrand as a two-term Taylor formula:

$$f(x, y) = f(x_i, y_i) + (x - x_i)[f_x(\xi, y(\xi)) + f_y(\xi, y(\xi)) f(\xi, y(\xi))]$$

where f_x and f_y denote the partial derivatives of f with respect to x and y. (Note that x is the independent variable and that y is a function of x. The quantity in square brackets is the total derivative of f with respect to x.) Now if we use the extended mean value theorem for integrals, we find

$$y_{i+1} = y_i + h f(x_i, y_i) + \int_{x_i}^{x_{i+1}} (x - x_i)(f_x + ff_y) \, dx$$

$$= y_i + h f(x_i, y_i) + \left(\frac{h^2}{2}\right)[f_x(\zeta, y(\zeta)) + f(\zeta, y(\zeta)) f_y(\zeta, y(\zeta))]$$

where $x_i < \zeta < x_{i+1}$.

The error term shows us that halving the step size reduces the local truncation error by a factor of about 4. However, now we need twice as many steps to get anywhere, so that the global reduction in error from halving the step size is about two. Figure 8.4 shows the effects of reducing the step size for a very

FIGURE 8.4

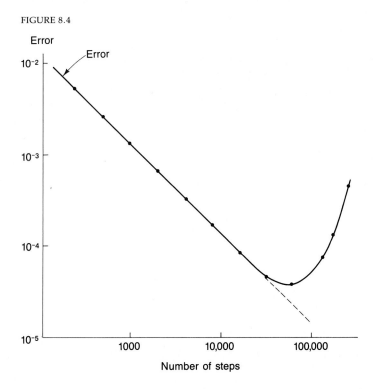

Number of steps

large number of steps. The initial-value problem was $dy/dx = y$, $y(0) = 1$ integrated from 0 to 1 using 32-bit arithmetic. The number of steps is the abscissa variable and the error in the solution at $x = 1$ is the ordinate variable on a log-log plot. The error decreases as predicted; the slope of the straight line in Figure 8.4 shows that the total error in determining $y(1.0)$ decreases by a factor of exactly 2 when the number of steps is doubled (step size halved). This behavior persists until the round-off error begins to have a substantial effect at around 50,000 steps. Such behavior limits the usefulness of Euler's method for high-accuracy solutions; the excessive number of steps required drives up the cost, and the accuracy gain in decreasing the step size is limited by the accumulation of round-off error. Nonetheless we study it as a simple example of how things work.

A somewhat more sophisticated method can be constructed by replacing $f(x, y)$ by a linear interpolation

$$f = \left[\frac{(x - x_i)}{h}\right] f(x_{i+1}, y_{i+1}) + \left[\frac{(x_{i+1} - x)}{h}\right] f(x_i, y_i)$$

If we do not mind for the moment that we do not have a value for y_{i+1}, then we can simply do the indicated integration to find

$$y_{i+1} = y_i + \left(\frac{h}{2}\right) [f(x_i, y_i) + f(x_{i+1}, y_{i+1})]$$

This is Heun's method.

It is usually hopeless to solve the Heun's method equation explicitly for y_{i+1}. Thus we use a different method, such as Euler's, to "predict" the value of y_{i+1} to use in the right-hand side of the Heun's method equation and use the Heun's method equation to find a more accurate value. This is a simple example of a group of methods known as *predictor-corrector* methods. Sometimes the corrector equation (in this case, Heun's equation) is used iteratively, but general practice in using predictor-corrector methods is to keep the step size small enough that the predictor (in this case, Euler's method) will give a value good enough that we need to use the corrector just once per step. We will see more of predictor-corrector methods later.

Example 8.2

Solve $dy/dx = xy$, $y(0) = 1$, using the Euler predictor and the Heun corrector with $h = 0.1$.

Predict: $f_0 = (0)(1) = 0$; $y_{1p} = y_0 + hf_0 = 1.0$

Correct: $y_{1c} = y_0 + \left(\frac{0.1}{2}\right)[0 + (0.1)(1.0)] = 1.005$

Predict: $f_1 = (0.1)(1.005) = 0.1005$; $y_{2p} = y_1 + hf_1 = 1.01505$

Correct: $y_{2c} = y_1 + \left(\frac{0.1}{2}\right)[0.1005 + (0.2)(1.01505)] = 1.0201755$

Recall that Euler's method gave $y_2 = 1.01$. The exact solution at $x = 0.2$ is 1.02020134, so we see that this method is considerably more accurate than Euler's method alone. □

The choice of $h = 0.1$ for the step size in this example was arbitrary. The choice of step size is an issue here (to be discussed later), just as the choice of panel width was an issue in the numerical evaluation of integrals. Before we get to that problem we need to derive methods that are more accurate than these.

8.2.1 More Sophisticated Methods: Multistep

The development of Euler's and Heun's methods is a simple introduction to a whole set of methods based upon the integral equation with which we dealt earlier in this section. We will generalize the procedure by introducing the Newton backward difference formula (NBF) to interpolate $f(x, y)$ in the integrand and classify the results into families depending upon the following considerations:

- How many terms in the NBF do we use? If the backward differences involve not only subscripts i but also $i - 1$, $i - 2$, . . . , then we generate what are called *multistep methods* because more than one previous solution point is involved in finding the new solution point.
- Do we simply integrate from x_i to x_{i+1} or do we integrate from x_{i-1} or x_{i-2} to x_{i+1}? We might feel that involving more past solution points in the integral could give us more accuracy.
- Do we use an NBF based upon x_i or upon x_{i+1}? If we use x_i as the basis for the NBF expression, then we generate what are called *open*, or *predictor-type*, formulas. If we use x_{i+1} as the basis for the NBF expression, then we generate what are called *closed*, or *corrector-type*, formulas.

Because this section (and Chapter 7) have dealt at some length with the details of such procedures, we will not perform detailed derivations of the formulas and their error terms. Some derivations will be suggested as problems in the problem sets.

The NBF

$$f = f_i + \alpha \nabla f_i + \left[\frac{\alpha(\alpha + 1)}{2}\right]\nabla^2 f_i + \left[\frac{\alpha(\alpha + 1)(\alpha + 2)}{3!}\right]\nabla^3 f_i + \cdots$$

is substituted for $f(x, y)$ in the integral. Here α is a local variable:

$$\alpha = \frac{x - x_i}{h}$$

In the integral the independent variable x is replaced by

$$x = x_i + \alpha h$$

where $h = x_{i+1} - x_i$, and dx is replaced by $h\, d\alpha$. The limits of integration then become 0 to 1, or -1 to 1, or -2 to 1, or -3 to 1, . . . , depending on how many steps we wish to cover in the integration. Shown next are four open formulas resulting from integration from x_i to x_{i+1}, or from 0 to 1 on α. These are known as *Adams-Bashforth formulas*. Note that the first one is also Euler's method.

Formula	**Error Term**
$y_{i+1} = y_i + hf_i$	$\left(\dfrac{1}{2}\right) h^2 f'(\zeta, y(\zeta))$
$y_{i+1} = y_i + \left(\dfrac{h}{2}\right)(3f_i - f_{i-1})$	$\left(\dfrac{5}{12}\right) h^3 f'' = (\zeta, y(\zeta))$
$y_{i+1} = y_i + \left(\dfrac{h}{12}\right)(23f_i - 16f_{i-1} + 5f_{i-2})$	$\left(\dfrac{3}{8}\right) h^4 f^{(3)}(\zeta, y(\zeta))$
$y_{i+1} = y_i + \left(\dfrac{h}{24}\right)(55f_i - 59f_{i-1} + 37f_{i-2} - 9f_{i-3})$	$\left(\dfrac{251}{720}\right) h^5 f^{(4)}(\zeta, y(\zeta))$

If we base our Newton backward interpolation formula on x_{i+1}, which reads

$$f = f_{i+1} + \alpha \nabla f_{i+1} + \left[\frac{\alpha(\alpha + 1)}{2}\right]\nabla^2 f_{i+1} + \cdots$$

then we obtain a different set of closed formulas, often called the *Adams-Moulton formulas*:

Formula	**Error Term**
$y_{i+1} = y_i + hf_{i+1}$	$-\left(\dfrac{1}{2}\right) h^2 f'(\zeta, y(\zeta))$
$y_{i+1} = y_i + \left(\dfrac{h}{2}\right)(f_{i+1} + f_i)$	$-\left(\dfrac{1}{12}\right) h^3 f''(\zeta, y(\zeta))$
$y_{i+1} = y_i + \left(\dfrac{h}{12}\right)(5f_{i+1} + 8f_i - f_{i-1})$	$-\left(\dfrac{1}{24}\right) h^4 f^{(3)}(\zeta, y(\zeta))$
$y_{i+1} = y_i + \left(\dfrac{h}{24}\right)(9f_{i+1} + 19f_i - 5f_{i-1} + f_{i-2})$	$-\left(\dfrac{19}{720}\right) h^5 f^{(4)}(\zeta, y(\zeta))$

Note that the second of these formulas is Heun's formula.

It is clear from comparing the error terms that, given near equality for the derivatives in the error terms, the errors are substantially lower for the Adams-Moulton formulas than for the corresponding Adams-Bashforth formulas. A pair of Adams formulas of the same order is often used as a predictor-corrector method; for example,

$$y_{i+1,p} = y_i + \left(\frac{h}{24}\right)(55f_i - 59f_{i-1} + 37f_{i-2} - 9f_{i-3}) \quad \text{(Adams-Bashforth)}$$

$$y_{i+1,c} = y_i + \left(\frac{h}{24}\right)(9f_{i+1} + 19f_i - 5f_{i-1} + f_{i-2}) \quad \text{(Adams-Moulton)}$$

where f_{i+1} in the corrector formula is evaluated using $y_{i+1,p}$ as the y value in f_{i+1}.

The reader may wonder how to use multistep methods, given that initial-value problems have only one starting value. This is, in fact, one of the disadvantages of multistep methods because they do need help in getting started. In the sample formulas just given, it is clear that three extra points on the solution curve must be found by some other method. A common way of obtaining the extra points is to use an accurate single-step method to generate the required extra points. We discuss such methods later in the chapter.

Other families of formulas can be generated by integrating not just from x_i to x_{i+1} but over more than one step. One such set of formulas popular many years ago is the Milne method. The predictor is obtained by integration of a second-order Newton formula from x_{i-3} to x_{i+1}:

$$y_{i+1,p} = y_{i-3} + \left(\frac{4h}{3}\right)(2f_i - f_{i-1} + 2f_{i-2})$$

The corrector formula is Simpson's rule:

$$y_{i+1,c} = y_{i-1} + \left(\frac{h}{3}\right)(f_{i+1} + 4f_i + f_{i-1})$$

Although this method was once widely used, it suffers from problems of numerical instability under certain conditions and is no longer recommended. The Adams methods are stable and are now the most widely used multistep methods for that reason. The subject of stability and what it means is treated in Section 8.6.

8.2.2   Multistep Methods: Error Estimation and Step-Size Control

Early in the chapter we mentioned the dependence of the local truncation error on the step size, but we deferred the matter of choice of step size until later. We now have some of the tools necessary to deal with this issue. As you might suspect, the choice of step size is important, but there is no easy, a priori method of choosing a value for h.

For predictor-corrector methods it is relatively easy to develop a method of controlling step size, provided that the predictor and corrector are chosen to be of the same order. (If they are not of the same order, the predictor may be too accurate to be efficient or not accurate enough to allow the corrector to give a good value in one iteration. Fortunately, the choice of formulas of the same order is usually a happy one.) We show how an error estimator can be constructed by means of an example.

Example 8.3

Derive an error estimator for the fourth-order Adams-Bashforth predictor and Adams-Moulton corrector.

Let y^* be the true value of $y(x_i)$. (Note that $y(x_i)$ does not, in general, equal y_i in our notation. y_i is the numerically computed value, whereas $y(x_i)$ is the value of the exact solution at $x = x_i$. They are known to be equal only for $i = 0$.) Also, let $y_{i+1,p}$ be the predictor value. Then we can write

$$y^* = y_{i+1,p} + (\tfrac{251}{720}) h^5 f^{(4)}(\xi) = y_p + E_p$$

Likewise, for the corrector we can write

$$y^* = y_{i+1,c} - (\tfrac{19}{720}) h^5 f^{(4)}(\zeta) = y_c + E_c$$

We know that although $\xi \neq \zeta$, the values of $f^{(4)}(\xi)$ and $f^{(4)}(\zeta)$ should be very nearly equal, so that

$$E_p \simeq - (\tfrac{251}{19}) E_c$$

Then we can write

$$y^* \simeq y_p - (\tfrac{251}{19}) E_c$$
$$y^* = y_c + E_c$$

and solve to find $E_c \simeq (\tfrac{19}{270})(y_p - y_c)$.   □

Once we have such an estimate, we can then set h such that E_c is acceptable. This can be done automatically by the integrator program, but that is not simple for multistep methods. Usually such programs provide for halving the value of h, which reduces the local truncation error by a factor of, for instance, 32 in the preceding example. Continuing the solution if the step size has been halved requires inserting new points among the previous points on the solution curve. This can be done either by accurate interpolation or by calling upon whatever starter program produced the extra initial points.

It is also possible that the step size should be increased. Changes in the behavior of the solution can require changes in the step size of either kind. We may wish the step size to be decreased to increase the accuracy, or we may wish the step size to be increased to reduce the computing time. It is necessary to be careful in setting the epsilons to trigger such changes. In the preceding example, unless the epsilon for initiating a step size increase is less than one thirty-second of the epsilon for initiating a step-size decrease, we run the risk of having the program ratchet the step size up and down.

In addition to step-size control, automatic multistep integrators sometimes incorporate variable order control. For instance, they might start with a first-order formula and very small steps and then switch to higher-order formulas in a sort of bootstrap process. Sophisticated multistep method programs to solve systems of ordinary differential equations have been written around such ideas [Shampine and Gordon, 1975].

8.3   Runge-Kutta Methods

Runge-Kutta methods are single-step methods, which means that only one solution point is involved in finding the next solution point. These methods use a weighted sum of the values of $f(x, y)$ evaluated at the starting point of each step and at various points across the integration step. If $f(x, y)$ is a complicated function, the computer time spent on function evaluations may be excessive. Nonetheless, these methods are stable, as we shall see later, and are relatively efficient if the function $f(x, y)$ is not too complicated. Furthermore, they have the advantage that a change of step size causes no problem.

The most-common fourth-order Runge-Kutta formula (attributed to Runge) for a single equation is

$$k_1 = f(x_i, y_i)$$

$$k_2 = f\left(x_i + \frac{h}{2}, y_i + \left(\frac{h}{2}\right)k_1\right)$$

$$k_3 = f\left(x_i + \frac{h}{2}, y_i + \left(\frac{h}{2}\right)k_2\right)$$

$$k_4 = f(x_i + h, y_i + hk_3)$$

$$y_{i+1} = y_i + \left(\frac{h}{6}\right)(k_1 + 2k_2 + 2k_3 + k_4)$$

The construction of such formulas is rather complicated, particularly for higher-order formulas, so we shall derive only a few second-order formulas to illustrate the method.

The essence of the construction is to assume a form of solution (for the second-order case)

$$y_{i+1} = y_i + h(Ak_1 + Bk_2)$$

with

$$k_1 = f(x_i, y_i)$$
$$k_2 = f(x_i + ah, y_i + bhk_1)$$

and where A, B, a, and b are arbitrary constants. We then expand y_{i+1} into a Taylor series in h and we also expand $y(x_{i+1})$ into a Taylor series. We then force the series terms to be equal through as many powers of h as possible. We will be able to make some arbitrary choices of the constants. We will assume that $y_i = y(x_i)$ but that $y_{i+1} \neq y(x_{i+1})$, in general.

First, we find that

$$y(x_i + h) = y(x_i) + hy'(x_i) + \frac{1}{2!}h^2 y''(x_i) + \frac{1}{3!}h^3 y'''(\xi)$$

$$= y(x_i) + hf(x_i, y_i) + \frac{1}{2!}h^2[f_x(x_i, y_i) + f_y(x_i, y_i)f(x_i, y_i)] + \frac{1}{3!}h^3 y'''(\xi)$$

$$x_i < \xi < x_{i+1}$$

where f_x means $\partial f/\partial x$ and f_y means $\partial f/\partial y$. Next we find

$$y_{i+1} = y_i + h[Ak_1 + Bk_2]$$

$$= y_i + Ahf(x_i, y_i) + Bhf(x_i + ah, y_i + bhk_1)$$

$$= y_i + Ahf(x_i, y_i) + Bh[f(x_i, y_i) + f_x(x_i, y_i)ah + f_y(x_i, y_i)f(x_i, y_i)bh + O(h^2)]$$

where $O(h^2)$ means that there are terms of degree 2 or higher in h. Now we equate coefficients of h^0, h^1, and h^2:

$$y_i = y(x_i)$$
$$(A + B)f(x_i, y_i) = f(x_i, y(x_i))$$
$$Baf_x(x_i, y_i) + Bbf_y(x_i, y_i)f(x_i, y_i) = \frac{1}{2}[f_x(x_i, y(x_i)) + f_y(x_i, y(x_i))f(x_i, y(x_i))]$$

From these we have the conditions

$$A + B = 1$$
$$aB = bB = \frac{1}{2}$$

We can create different methods because we have only three conditions for four coefficients. The choices for the four coefficients have names associated with them.

1. $B = 1$, $A = 0$, $a = b = \frac{1}{2}$. Sometimes known as the modified Euler's method, this procedure consists of a two-step process:

$$y_{i+1/2} = y_i + \left(\frac{h}{2}\right)f(x_i, y_i)$$

$$y_{i+1} = y_i + hf(x_i + \frac{h}{2}, y_{i+1/2})$$

The method consists of an Euler-type half-step to the middle of the interval. At this point the slope is evaluated and used in an Euler whole step. It gives somewhat better accuracy than Euler's original method because the slope at the (estimated) midpoint is more representative of the whole interval slope than the slope at the left end.

2. $A = B = \frac{1}{2}$, $a = b = 1$. This is Heun's method, which we have seen before. It also proceeds in two stages:

$$y_{i+1} = y_i + hf(x_i, y_i)$$
$$y_{i+1} = y_i + \left(\frac{h}{2}\right)[f(x_i, y_i) + f(x_{i+1}, y_{i+1})]$$

This method uses an Euler estimate for y_{i+1} in order to find the approximate slope at x_{i+1}, and is, in reality, a one-step predictor-corrector method. The slopes at the beginning and the end of the interval are averaged in order to approximate the average slope across the interval.

Higher-order Runge-Kutta methods are more common than the lower-order ones, and the fourth-order families are some of the most popular. Another method, attributed to Kutta, is the following:

$$y_{i+1} = y_i + \left(\frac{h}{8}\right)[k_1 + 3k_2 + 3k_3 + k_4]$$

where

$$k_1 = f(x_i, y_i)$$

$$k_2 = f(x_i + \frac{1}{3}h, y_i + \frac{1}{3}hk_i)$$

$$k_3 = f(x_i + \frac{2}{3}h, y_i - \left(\frac{1}{3h}\right)k_1 + hk_2)$$

$$k_4 = f(x_i + h, y_i + hk_1 - hk_2 + hk_3)$$

Example 8.4

Integrate $dy/dx = xy$, $y(0) = 1.0$, using the Runge version of Runge-Kutta formulas with a step size of 0.1.

$$k_1 = (0)(1) = 0.0$$
$$k_2 = (0 + 0.05)[1 + 0.05(0)] = 0.05$$
$$k_3 = (0 + 0.05)[1 + 0.05(0.05)] = 0.050125$$
$$k_4 = (0 + 0.10)[1 + 0.1(0.050125)] = 0.10050125$$

$$y(0.1) = 1.0 + \left(\frac{0.1}{6}\right)[0.0 + 2(0.05) + 2(0.050125) + 0.10050125]$$

$$= 1.00501\ 2521 \qquad \text{(versus } 1.00501\ 2521 \text{ true value)}$$

$$k_1 = (0.1)(1.00501\ 2521) = 0.10050\ 12521$$
$$k_2 = (0.15)[1.005\ldots + 0.05(0.10050\ 12521)] = 0.15150\ 56375$$
$$k_3 = (0.15)[1.005\ldots + 0.05(0.15150\ 56375)] = 0.15188\ 81704$$
$$k_4 = (0.20)[1.005\ldots + 0.10(0.15188\ 81704)] = 0.20404\ 02676$$

$$y(0.2) = 1.00501\ 2521 + \left(\frac{0.1}{6}\right)[0.10050\ 12521 + 2(0.15150\ 56375)$$

$$+ 2(0.15188\ 81704) + 0.20404\ 02676]$$

$$= 1.02020\ 1340 \qquad \text{(versus } 1.02020\ 1340 \text{ true value)}$$

We see that we get much more accuracy than we got with Euler's method for the same step size but at the expense of much more computation. □

There are other Runge-Kutta varieties with properties that are determined by the choices of the arbitrary constants. The PASCAL program RKF45, which we will see later in the chapter, uses fourth- and fifth-order Runge-Kutta formulas to integrate each step. The fifth-order method is seldom used by itself because it requires six function evaluations per step, but in this case the constants are cleverly chosen so that the same evaluation points can be used in both formulas. Thus only six function evaluations instead of ten are used. The results of the two different methods are compared in order to provide information for step-size control. The IMSL program DVERK is similar in nature, but it uses fifth- and sixth-order Runge-Kutta methods.

8.4   Comparison of Methods

In choosing among methods the user must be conscious of several important considerations. One factor in the choice of method can be the complexity of $f(x, y)$. If it is a complicated function, then the user of a Runge-Kutta method may pay a heavy penalty because of the number of function evaluations per integration step. Multistep predictor-corrector methods require only two evaluations per step unless the corrector is iterated.

In most good automatic integrators of any type, the step size is chosen by the program, although in some cases the subroutine may require an initial estimated h, which then may be altered by the integrator. Some programs even allow the user to make the choice of step size by overriding the automatic change feature.

Another consideration is the matter of "stiffness" of the differential equation. This property is one that we will look at later, but suffice it to say for now that stiff equations require special methods. Most integrators either will report failure or will require excessive computer time for stiff equations.

Finally, the user needs to be aware of the accuracy required for the solution. RKF45 and DVERK are moderately accurate. The IMSL routine DGEAR offers a choice of a higher-order (high-accuracy) Adams method or of a "backward" method for stiff equations. Most programs have some indication of the error tolerances that will be accepted by that program and may fail (or refuse to operate) if more accuracy is requested than can be satisfied.

8.5   Solving Systems of Differential Equations Numerically

Systems of differential equations are solved using the same methods we have developed, but these methods are now adapted to a set of equations. Our first example is the solution of the system

$$\frac{dx}{dt} = xy + 1 \qquad x(0) = 1$$

$$\frac{dy}{dt} = x + y \qquad y(0) = 0$$

using Euler's method for simplicity. The Euler equations become

$$x_{i+1} = x_i + h(x_i y_i + 1) \qquad x_0 = 1$$
$$y_{i+1} = y_i + h(x_i + y_i) \qquad y_0 = 0$$

If we let $h = 0.1$, then we have

$$x_1 = 1 + 0.1[(0)(1) + 1] = 1.1$$

$$y_1 = 0 + 0.1(1 + 0) = 0.1$$

$$x_2 = 1.1 + 0.1[(1.1)(0.1) + 1] = 1.211$$

$$y_2 = 0.1 + 0.1(1.1 + 0.1) = 0.22$$

If we do the same problem by the Runge-Kutta fourth-order method, then we need to introduce a set of k's with double subscripts: k_{ji}, where j refers to the particular differential equation and i refers to the number of the stage in the Runge-Kutta process. If we are using the Runge-Kutta method on a set of equations, it is important that we complete each stage of the process on all equations before we move on to the next stage. Thus

$$\left.\begin{array}{l} k_{11} = (1)(0) + 1 = 1 \\ k_{21} = 1 + 0 = 1 \end{array}\right\} \text{Stage 1}$$

$$\left.\begin{array}{l} k_{12} = [1 + 0.05(1)][0 + 0.05(1)] + 1 = 1.0525 \\ k_{22} = [1 + 0.05(1)] + [0 + 0.05(1)] = 1.10 \end{array}\right\} \text{Stage 2}$$

$$\left.\begin{array}{l} k_{13} = [1 + 0.05(1.0525)][0 + 0.05(1.10)] + 1 = 1.05789\,4375 \\ k_{23} = [1 + 0.05(1.0525)] + [0 + 0.05(1.10)] = 1.107625 \end{array}\right\} \text{Stage 3}$$

$$\left.\begin{array}{l} k_{14} = (1 + 0.10(1.05789\,4375)[(0 + 0.10(1.107625)] + 1 \\ \quad = 1.12248\,0003 \\ k_{24} = [1 + 0.10(1.05789\,4375)] + [0 + 0.10(1.107625)] \\ \quad = 1.21655\,1938 \end{array}\right\} \text{Stage 4}$$

$$x(0.1) = 1 + \left(\frac{0.1}{6}\right)(1 + 2(1.0525) + 2(1.05789\,4375)$$
$$\qquad + 1.12248\,0003) = 1.10572\,1146$$

$$y(0.1) = 0 + \left(\frac{0.1}{6}\right)[1 + 2(1.10) + 2(1.107625) + 1.21655\,1938]$$
$$\qquad = 0.11053\,00323$$

As a final example, we shall look at the initial value problem

$$\frac{d^2y}{dt^2} + 4y = 0, \qquad y(0) = 0, \qquad y'(0) = 2$$

We know (or can show) that the solution of this problem is $y(t) = \sin 2t$. We will use the third-order Adams-Bashforth method as a predictor:

$$y_{i+1,p} = y_i + \left(\frac{h}{12}\right)(23f_i - 16f_{i-1} + 5f_{i-2})$$

and the third-order Adams-Moulton method as a corrector:

$$y_{i+1,c} = y_i + \left(\frac{h}{12}\right)(5f_{i+1} + 8f_i - f_{i-1})$$

We see that in order to find y_3 we need y_2, y_1, and y_0, or three starting values. Since the point of this example is to show how the predictor-corrector method works, we will generate starting values from the known solution.

First we must produce the system by letting $x = dy/dt$, so that the initial value problem becomes, in standard form,

$$\frac{dx}{dt} = -4y, \qquad x(0) = 2$$

$$\frac{dy}{dt} = x, \qquad y(0) = 0$$

Now if $y = \sin 2t$, then $y(0) = 0$, $y(0.1) = 0.198669308$, $y(0.2) = 0.3894183423$. Because $x = y' = 2\cos 2t$, we can also write $x(0) = 2$, $x(0.1) = 1.960133156$, and $x(0.2) = 1.842121988$.

If $h = 0.1$, we can now find the predictor values:

$$y_{3,p} = y(0.2) + \left(\frac{h}{12}\right)[23(0.2) - 16(0.1) + 5(0)]$$

$$= 0.3894183423 + \left(\frac{0.1}{12}\right)[(23)(1.842121988)$$

$$- (16)(1.960133156) + (5)(2)]$$

$$= 0.3894183423 + 0.1750556269 = 0.5644739692$$

$$x_{3,p} = x(0.2) + \left(\frac{h}{12}\right)[23(-4)y(0.2) - 16(-4)y(0.1) + 5(-4)y(0)]$$

$$= 1.842121988 + \left(\frac{0.1}{12}\right)[23(-4)(0.3894183423)$$

$$- 16(-4)(0.198669308) + 5(-4)(0)]$$

$$= 1.842121988 - 0.1925970982 = 1.64952489$$

Next we use the correctors:

$$y_{3,c} = y(0.2) + \left(\frac{h}{12}\right)(5x(0.3) + 8x(0.2) - x(0.1))$$

$$= 0.3894183423 + \left(\frac{0.1}{12}\right)[5(1.64952489) + 8(1.842121988)$$

$$- 1.960133156]$$

$$= 0.3894183423 + 0.1752038933 = 0.5646222356$$

$$x_{3,c} = x(0.2) + \left(\frac{h}{12}\right)[5(-4)y(0.3) + 8(-4)y(0.2) - (-4)y(0.1)]$$

$$= 1.84212\,1988 + \left(\frac{0.1}{12}\right)[5(-4)(0.56462\,22356)$$

$$+ 8(-4)(0.38941\,83423) - (-4)(0.19866\,9308)]$$

$$= 1.84212\,1988 - 0.19132\,62869 = 1.65079\,5701$$

The exact values to 10 decimal places are $y(0.3) = 0.5646424734$ and $x(0.3) = 1.650671230$. Better accuracy would have been attained with a higher-order formula or smaller step size (or both).

8.6 Stability

Two different kinds of stability problems are encountered in solving differential equations numerically. One is the possible inherent instability of the differential equation solution regardless of the method of solution, and the other is the possible instability of the numerical solution method regardless of the stability of the mathematical solution. An example of the first kind of instability is the solution of the initial-value problem

$$\frac{d^2y}{dx^2} - 4\frac{dy}{dx} - 5y = 0$$

$$y(0) = 1, y'(0) = -1$$

The general solution of the differential equation is

$$y = c_1e^{-x} + c_2e^{5x}$$

The initial values specify the solution of the initial-value problem as

$$y = e^{-x}$$

Now if we perturb the initial condition $y(0) = 1$ to become $y(0) = 1 + \epsilon$, then we find that the solution to the new initial-value problem is

$$y = \left(1 + \frac{5\epsilon}{6}\right)e^{-x} + \frac{\epsilon}{6}\,e^{5x}$$

which is shown in Figure 8.5 for several very small values of ϵ. Any attempt to solve the original initial-value problem numerically is likely to produce bad results because round-off errors will start up the solution e^{5x}, which will eventually overwhelm the desired solution. This kind of instability is akin to ill-conditioning in systems of linear algebraic equations. Small perturbations of any of the coefficients of ill-conditioned linear-equation systems can cause drastic changes in the solutions. Likewise, for unstable differential equations, small perturbations of the initial conditions can cause large changes in the solutions. Frequently, the

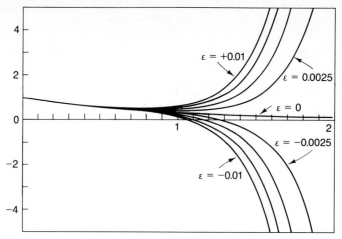

FIGURE 8.5   Solutions of $\dfrac{d^2y}{dx^2} - 4\dfrac{dy}{dx} - 5y = 0$, $y(0) = 1 + \epsilon$, $y'(0) = -1$ for various values of ϵ.

existence of such unstable solutions in the equations for some physical system indicates trouble in the design.

The kind of instability with which we are more concerned here results from the replacement of a first-order ordinary differential equation by a higher-order difference equation. The difference equation, if it is of order higher than first, will have multiple solutions. One of the solutions will correspond to the solution of the differential equation, but the others, often called extraneous solutions, may cause trouble just as the solution e^{5x} caused trouble in the initial-value problem example. The design of good numerical methods for integrating differential equations requires that the extraneous solutions be of the sort that "damp out," or decrease in magnitude, as the solution progresses. A complete discussion of stability is beyond the level of this text, but we will look at some of the essential points with a view toward understanding why some methods are stable and others are not.

The first condition that we will require of an integrator is that it be able to integrate $y'(x) = 0$, $y(0) = 1$, correctly. If we have a general integration method of the form

$$y_{i+1} = \sum_{j=i-m}^{j=i} a_j y_j + \sum_{j=i-n}^{j=i+1} b_j f(x_j, y_j)$$

where $m \geqslant 0$ and $n \geqslant 0$, then it follows that

$$\sum_{j=i-m}^{j=i} a_j = 1$$

in order to give an accurate solution. This property is called *consistency*.

Now suppose that we apply this method to $y' = 0$. Then we have a difference equation of the form

$$y_{i+1} = a_i y_i + a_{i-1} y_{i-1} + \cdots + a_{i-m} y_{i-m}$$

The solution of such a linear difference equation is found by setting

$$y_i = cs^i$$

where c is an arbitrary constant. The result of this substitution is a polynomial equation in s of the form

$$s^{i+1} - a_i s^i - a_{i-1} s^{i-1} - \cdots - a_{i-m} s^{i-m} = 0$$

When we have solved this for the roots s, we can write a solution for y_i as a linear combination of the various solutions for s in a way reminiscent of the procedure in linear differential equations:

$$y_i = c_1 s_1^i + c_2 s_2^i + \cdots$$

for all nonzero roots s. If this is to give $y_i = 1$ properly, then one of the roots must be $s = 1$. If any $|s| > 1$, then round-off error can cause the solution to grow without bound, which is, in this instance, undesirable. Any roots for which $|s| < 1$ correspond to extraneous solutions, which, when excited by round-off, will die out with increasing i. A general theorem on the roots of the characteristic polynomial equation is as follows: An integrator is stable if all the roots satisfy $|s| \leq 1$, and any roots for which $|s| = 1$ must be simple (e.g., $s = 1, 1$ would not be permitted, but $s = 1, -1$ would). If all roots but one satisfy $|s| < 1$ and the one root is $|s| = 1$, then the method is strongly stable. An exercise is to apply this to the Milne corrector to show that it is stable but not strongly stable.

If we look at the more general case of $y' \neq 0$, then the subject of stability is more complicated. We refer the reader to Atkinson [1978], where the subject is treated in detail. The matter of stability regions becomes important, which may put restrictions on the step size of integrator in relation to other parameters of the differential equation.

8.7 Stiff Differential Equations

A differential equation or system of differential equations having solutions with two or more widely varying time constants is known as a *stiff differential equation* or system. A time constant in such a system is the time that it takes for a transient, or short-lived, part of a solution to die away to $1/e$ of its initial value.

The numerical solution of a stiff equation or system is very difficult. In order to get an accurate solution of the short time constant part of the solution, we must use very small steps to get good accuracy. Once this part of the solution is done, we are tempted to lengthen the solution steps because the remaining part of the solution is very slowly changing and should yield an accurate solution with relatively long steps. Unfortunately, although the transient part of the solution has long ago become insignificant, the differential equation has not changed, and large steps bring on the same instability that we have seen before.

Very small steps throughout a large range of the independent variable can be very costly in terms of computer time. One way of avoiding small steps is to resort to what is called *backward integration* to solve the step-size problem after we have passed beyond the fast transient region. The simplest backward method is the backward Euler method, which evaluates the slope of the curve at the end of the integration step rather than at the beginning. Because we do not know y_{i+1}, we cannot use it in the right-hand side except implicitly, which means that we must be able to solve the integration equation as a nonlinear equation for the value of y_{i+1}. The backward Euler equation (the lowest-order Adams-Moulton method), for example, is

$$y_{i+1} = y_i + hf(x_{i+1}, y_{i+1})$$

In all but the simplest cases (linear equations, which should be solved by other means) this leads to a nonlinear equation or system of equations that requires an iterative solution, say by Newton-Raphson methods. This is also an expensive process, and it is very difficult to say in many situations whether it is more efficient to put up with small step sizes or to put up with the nonlinear equation solutions. The only rule that seems to apply is that the stiffer the equation is, the more attractive the backward integration method seems to be.

To show why backward integration works in such cases, we can take a very simple stiff equation

$$\frac{dy}{dt} = -ay + f(t), \qquad y(0) = y_0$$

where a is a very large positive number and $f(t)$ is a relatively slowly changing function. The classical solution for this equation is

$$y(t) = y_0 e^{-at} + e^{-at} \int_0^t e^{ax} f(x)\, dx$$

If we choose $f(t) = 1$, a *very* slowly changing function, the solution becomes

$$y(t) = \left(y_0 - \frac{1}{a} \right) e^{-at} + \frac{1}{a}$$

If $a = 100$, for example, the transient part of the solution will have vanished for all practical purposes in much less than 1 s. If we apply a simple numerical method (Euler's method) to the same equation, we have

$$y_{i+1} = y_i - ahy_i + h$$

This difference equation has the particular solution $y_i = 1/a$; the complete solution is

$$y_i = \left(y_0 - \frac{1}{a} \right)(1 - ah)^i + \frac{1}{a}$$

This is stable as long as $0 < 1 - ah < 1$, which means that if a is very large, then h must be very small even after the term e^{-at} has ceased to contribute. If h is ever increased so that $1 - ah < -1$, the solution will "blow up."

If we use the Euler backward equation instead of Euler's forward equation, we obtain

$$y_{i+1} = y_i - ahy_{i+1} + h$$

which has the complete solution

$$y_i = \frac{y_0 - 1/a}{(1 + ah)^i} + \frac{1}{a}$$

Note that here the size of h can be increased after e^{-at} is negligible without causing a stability problem.

Why could we not have thrown away the offending term after it had outlived its usefulness? In this simple example, we could have done so. However, in most cases that is not possible, and we are forced to live with the problem.

8.8 Boundary-Value Problems

A two-point boundary-value problem is a second-order differential equation system in which we do not have two initial conditions given. Instead, we have a condition on the solution at two different points bounding an interval over which we would like the solution. These problems come up frequently in the solution of partial differential equations, where we can separate variables and create two or more ordinary differential equation problems. An example is

$$\frac{d^2y}{dx^2} + \frac{dy}{dx} + y = 0, \qquad y(0) = 0, \qquad y(1) = 1$$

in which we want the solution over $(0, 1)$. Since this is not a standard initial-value problem, we cannot directly apply the methods we have learned without modification.

8.8.1 The Shooting Method

In the first method we use standard numerical integration techniques after we have modified the problem to an initial-value problem. We replace the condition at one end of the boundary by a trial slope condition at the other end. Thus the name: We shoot at the target boundary value at the far end. In the example just given, we make a guess at the slope at $x = 0$, y_0', and convert the equations and conditions to

$$\frac{dv}{dx} = -v - y$$

$$\frac{dy}{dx} = v$$

$$y(0) = 0$$

$$v(0) = y_0'$$

This can be solved over (0, 1) by any standard integration method. If we have missed the target—that is, if $y(1) \neq 1$—we adjust y_0' up if $y(1) < 1$ or down if $y(1) > 1$ and try again. Once we have two incorrect estimates, we can start using linear interpolation to improve on the estimates.

Example 8.5

An Euler initial-value solver with steps of $h = 0.01$ was used on the example problem with initial guesses for y_0' of 1 and 2. Given is a listing of the results. The interpolations were done by hand (in fact, the first interpolation, 1.9, was a quick guess).

Initial Slope	Y(1)
1.0	0.536825
2.0	1.073650
1.9	1.019967
1.873	1.005473
1.8656	1.001500 □

The advantage of the shooting method is that it uses software that is (usually) readily available. The interpolation can be done in the machine, but there is a good deal to be said for using the system interactively and doing the interpolation by calculator. Sometimes that leads to increased understanding of the problem.

It is necessary to recognize that some such problems do not have a unique solution. Firing a gun at a target within range of the gun is an example. We can aim in the usual way, or we can fire the bullet in a high trajectory (almost straight up if the target is very close). The latter is not a recommended method, but it is a solution to the problem. Sometimes one of the boundary conditions may be a derivative condition. In that case, if the shooting is to be done from the end for which y is fixed, then the "target" is the correct slope at the other end. If the shooting is to be done from the end that has the slope condition, then a value of y has to be assumed. In neither case does this really complicate the solution.

8.8.2 Finite-difference Method

Another way to solve a boundary-value problem such as the preceding example is to replace the derivatives by divided differences:

$$y' \rightarrow \frac{y(x_{i+1}) - y(x_i)}{h}$$

$$y'' \rightarrow \frac{y(x_{i+1}) - 2y(x_i) + y(x_{i-1})}{h^2}$$

Then the example problem becomes

$$y(x_{i+1})(1 - h) + y(x_i)(h^2 - h - 2) + y(x_{i-1}) = 0 \qquad i = 1, 2, \ldots, n-1$$

with $y(x_0) = 0$ and $y(x_n) = 1$. This is a tridiagonal system, which can be solved by the methods discussed in Chapter 4. If h is not large the system is diagonally dominant and can be solved without worry about pivoting.

However, the example is a linear equation, which can be solved classically. What if the differential equation is nonlinear? Unfortunately, we then have a set of nonlinear difference equations, which may be very difficult to solve, as most sets of nonlinear equations are. Chapter 5 does have some discussion of nonlinear systems, but an intensive treatment is beyond the scope of this book.

8.9   Procedure RKF45

RKF45 is a Runge-Kutta-Fehlberg combination fourth- and fifth-order Runge-Kutta–type single-step differential equation solver. It is useful for the solution of nonstiff or mildly stiff systems of ordinary differential equations in which extreme accuracy is not needed. Because most "casual" problems probably fall into this category, it is a good solver to try first if nothing is known about the differential equation or system. It is adaptive in the sense that it adjusts step size (within limits) to try to meet user-set criteria on the solution error.

The call to procedure RKF45 is

```
RKF45 (NEQN, Y0, Y, TIN, TOUT, RELERR, ABSERR, FLAG);
```

where

NEQN	=	the number of simultaneous first-order ODEs
Y0	=	the vector of initial values of the dependent variables upon call
Y	=	upon return, the solution vector
TIN	=	starting value of the independent variable
TOUT	=	ending point of the integration
RELERR	=	relative error criterion set by the user
ABSERR	=	absolute error criterion set by the user
FLAG	=	integer communication flag

RELERR and ABSERR are nonnegative error tolerances. RELERR should never be zero, but ABSERR may be zero. If the solution goes to zero, the relative error test is meaningless and the procedure will be aborted with FLAG = 4 if ABSERR is also zero. Although the original versions of RKF45 recommend that RELERR not be set less than 10^{-8}, it does work for some problems with RELERR set 100 or more times smaller.

FLAG is used in communicating information in both directions. On the calls to RKF45, FLAG should be set to +1 for integration from TIN to TOUT. FLAG = −1 will cause the integrator to make one RKF (internal) step from TIN toward TOUT. Upon return, FLAG = 2 for normal termination of a FLAG = 1 call. The exit from a single-step (FLAG = −1) will result in FLAG = −2 upon normal termination. At each call FLAG should be set to +1 or −1. In the single-step case, the integrator will stop when TOUT is reached. Upon return, FLAG = 3 indicates that an excessive number of function evaluations was required. FLAG = 4 indicates that the pure relative error test failed and ABSERR should be set to a nonzero value. FLAG = 5 indicates that the requested error tolerances could not be met with the smallest step size. The error tolerances should be increased. FLAG = 6 indicates that too many output points have been requested. Consider using a different method or using the single-step mode. FLAG = 7 indicates that input parameters or FLAG values are wrong.

The user must provide a procedure to define the differential equations. The header for DIFFEQ should be

```
PROCEDURE DIFFEQ(T: REAL; VAR Y, YP: VECTOR);
```

The values of the derivatives are passed back to RKF45 in the vector YP. An example shows how the procedures are used.

Example 8.6

We wish to solve the initial-value problem

$$y'' = -4y, \qquad y(0) = 0, \qquad y'(0) = 6$$

(The solution is $y = 3 \sin 2t$.) First we let $y' = z$, so that we have the system

$$y' = z, \qquad y(0) = 0$$
$$z' = -4y, \qquad z(0) = 6$$

Let y be Y[1], z be Y[2], y' be YP[1], and z' be YP[2]. Our procedure then should be

```
PROCEDURE DIFFEQ(T: REAL; VAR Y, YP: VECTOR);
BEGIN
   YP[1] := Y[2];
   YP[2] := -4.0*Y[1]
END;
```

The main program must contain the following declaration:

```
CONST NEQN = 2;
TYPE VECTOR = ARRAY[1..NEQN] OF REAL;
VAR
                    U26, TIN, TOUT: REAL;
                      Y, Y0, YP: VECTOR;
        RELERR, ABSERR, OLDRELERR, OLDABSERR: REAL;
                 FLAG, EXITFLAG, OLDFLAG: INTEGER;
                               INIT: BOOLEAN;
```

One of the first executable statements must be FINDU26, which calls a procedure that calculates 13 times the machine ϵ used in error control. If we wish to obtain the solution of the differential equation system from $t = 0$ to $t = 3$ at every $\frac{1}{4}$ unit point, then we need statements in the main program that include the following:

```
FINDU26;
Y0[1] :=0; Y0[2] := 6.0;
ABSERR := . . . ; RELERR := ...; (10⁻⁹ suitable)
TIN := 0; TOUT := 0.25; FLAG := 1;
(Print initial values . . . )
REPEAT
    RKF45(NEQN, TIN, TOUT, Y0, Y, RELERR, ABSERR, FLAG);
    (Print results . . . )
    Y0 := Y; FLAG := 1; TOUT := TOUT + 0.25
UNTIL TOUT > 3.1;
```

The results of this calculation are shown for 64-bit arithmetic floating-point hardware with RELERR = ABSERR = 10^{-9}. Note that these values are good to about nine decimal places.

y	t
0.00000 00000 0000E+000	0.00
1.43827 66165 8794E+000	0.25
2.52441 29566 7155E+000	0.50
2.99248 49631 2952E+000	0.75
2.72789 22840 8756E+000	1.00
1.79541 64352 2118E+000	1.25
4.23360 02538 1014E-001	1.50
-1.05234 96855 3592E+000	1.75
-2.27040 74927 5821E+000	2.00
-2.93259 03626 9700E+000	2.25
-2.87677 28342 8467E+000	2.50
-2.11662 09849 7389E+000	2.75
-8.38246 49864 1547E-001	3.00 □

8.10   IMSL Subroutine DVERK

DVERK is a Runge-Kutta fifth- and sixth-order subroutine for solving a system of first-order ordinary differential equations. It is good for nonstiff equation systems with relatively simple functions for the derivatives. It is expensive to use with graphical solutions that require many closely spaced solution points.

The call to DVERK is

CALL DVERK (N, FCN, X, Y, XEND, TOL, IND, C, NW, W, IER)

where the parameters are

N	=	number of equations in the system, input
FCN	=	name of subroutine for evaluating derivatives
X	=	independent variable. On input X is the initial value and on output X = XEND if step was successful
Y	=	vector of length N containing the initial values of the dependent variables on input and the computed values at X = XEND on successful output
XEND	=	value of X at which the solution is desired
TOL	=	error control tolerance (If IND = 1, DVERK attempts to control the error in such a way as to limit the maximum over the set of ODEs of the estimated relative error of the solution to the value of TOL. If the absolute value of the solution drops below 1.0, the absolute error is used instead of the relative error.)
IND	=	indicator, used on input and output (See later material for options.)
C	=	communication vector of length 24 (If IND = 1 (easy solution method) the contents of C are of no concern to the user. See the following.)
NW	=	number of rows in the dimension statement for the work matrix W; NW must be N or greater
W	=	work matrix dimensioned with NW rows and at least 9 columns; contents of W must not be modified by the user
IER	=	error parameter, output

  IER = 129   NW < N or TOL <= 0
  IER = 130   IND not in range from 1 to 6
  IER = 131   XEND has not been changed from previous call or X is not set to previous XEND value
  IER = 132   relative error problem (IND = 2 case)

The easy way to use DVERK is to set IND = 1, in which case setting the components of the C vector is unnecessary. IND = 2 allows the user a range of options regarding error and step-size control not available under the IND = 1 option. The variety of these options is more than it is appropriate to present here; the user should read the IMSL manual carefully in order to understand these options. The IND = 1 easy option is sufficient for most problems.

IND also acts as an error indicator. IND = -3 on return signifies that the integrator was not able to satisfy the error tolerance, probably because TOL was too small. IND = -2 also indicates a problem with the step size, again probably because TOL was too small. IND = -1 indicates that the number of function evaluations was excessive, but this happens only under the IND = 2 input option, when a limit has been placed on the number of function evaluations in C(7).

To use DVERK one must write a subroutine whose header is of the form

```
SUBROUTINE name (N, X, Y, YP)
```

where N is the number of equations, X is the independent variable, Y is the vector of values of the independent variables, and YP is the vector of values of the derivatives. For the fox-and-rabbit problem at the beginning of the chapter, in which

$$\frac{df}{dt} = -pf + qfr$$

$$\frac{dr}{dt} = -arf + br$$

we could consider f as Y(1) and r as Y(2). Then the complete subroutine would be

```
SUBROUTINE name (N, X, Y, YP)
INTEGER N
REAL YP (N) , Y (N) , X
DATA A, B, P, Q/ . . . values . . . /
YP (1) = -P*Y (1) + Q*Y (1) *Y (2)
YP (2) = -A*Y (2) *Y (1) + B*Y (2)
RETURN
END
```

The name of this subroutine must be declared EXTERNAL in the main program so that the subroutine name can be passed as a parameter.

In the main program the call to DVERK requires (in the easy-to-use IND = 1 mode) the setting of N (N = 2 for the fox-and-rabbit problem), the error tolerance TOL, the name of the subroutine, the names of the communication vector and the work matrix (as well as its row dimension), and IND = 1. The value of X is set at its initial value and the vector Y is filled (in proper order) with the initial values of the dependent variables. XEND is set to the first value of X for which the solution is desired. Upon successful return, X is now equal to XEND and Y contains the calculated values of the dependent variables. To continue the solution, it is necessary only to change the value of XEND to the desired next value of the independent variable. The values in vector Y are now initial values for the next step. Of course, it is wise to check that IER is zero and that IND is not negative as a result of the previous step.

Exercises

1. Show that if $f(x, y)$ is a well-behaved function in a region, then solutions of $dy/dx = f(x, y)$ cannot cross each other.

2. Derive the corrector error estimates for the per-step error for each of the first three

predictor-corrector formulas constructed by using the Adams-Bashforth predictor and the corresponding Adams-Moulton corrector.

3. Find the general solution of the differential equation

$$\frac{dy}{dx} = xy(y - 1)$$

by separation of variables. Also show that if $y(0) = 1$, then the solution reduces to $y(x) = 1$. Now find the solution for the initial condition $y(0) = 1 + t$ and graph the solutions for $t = 0$, $t < 0$, and $t > 0$, with t small. Comment on the prospects of producing an accurate solution for $y(0) = 1.001$.

4. Derive the Euler method of solution by writing $y(x)$ as a Taylor series expansion and truncating the series. Revise the derivation to provide a truncation error estimate.

5. Derive the Euler method of solution by a geometric construction. The point of view should be that of extrapolating the solution linearly with a tangent line from the previous solution point (or the initial point).

For each of the following differential equations, find the solution over the interval $[0, 5]$ unless otherwise indicated.

6. $d^3y/dx^3 + 5\,d^2y/dx^2 + 9\,dy/dx + 5y = 3e^{2x}$, $y(0) = 3$, $y'(0) = 0$, $y''(0) = 0$

7. $d^2y/dx^2 + 2y - 2y^3 = 0$, $y(0) = 0$, $y'(0) = 1$

8. $2yy'' - (y')^2 = 1$, $y(0) = 2$, $y'(0) = -1$

9. $d^2y/dx^2 = 2y^3$, $y(0) = -\frac{1}{3}$, $y'(0) = -\frac{1}{9}$
   (Solution only for $0 \leqslant x < 3$)

10. $x^4\,d^3y/dx^3 - 12y^4 = 0$, $y(0) = 0$, $y'(0) = \frac{1}{2}$, $y''(0) = -\frac{1}{2}$

11. Derive one or more of the Adams-Bashforth or Adams-Moulton integrators shown in Section 8.2.1.

12. (a) Investigate the stability of the Milne method corrector equation by finding the zeros of the characteristic polynomial.
    (b) Find the difference equation that results from using the Milne corrector on the simple equation $y' = Ay$, where A is a constant. Show that although the corrector is stable for $A > 0$ (the extraneous solution damps out), it is unstable for $A < 0$ (the extraneous solution grows in magnitude).

13. Hamming [1959] proposed a multistep predictor-corrector method, which uses the Milne predictor:

$$y_{i+1,p} = y_i + \left(\frac{4h}{3}\right)(2f_i - f_{i-1} + 2f_{i-2})$$

Next he assumed that things do not change much from one step to the next and proposed a modifier:

$$y_{i+1,m} = y_{i+1,p} + \frac{112}{121}(y_{i,c} - y_{i,p})$$

which would give a better value to use initially in the corrector in the function f_{i+1}:

$$y_{i+1,c} = \frac{9y_{i,c} - y_{i-2,c}}{8} + \left(\frac{3h}{8}\right)(f_{i+1} + 2f_i - f_{i-1})$$

What are the roots of the characteristic polynomial equation? What can you say about the stability of this method?

Problems

PROJECT PROBLEM 8.1

Solve the predator-prey set of differential equations from the example at the beginning of the chapter. Produce graphs of the fox and rabbit populations over several cycles.

For the same set of initial conditions, double the birth rate of the rabbits and observe several cycles of the population curves.

PROJECT PROBLEM 8.2

The Wabash Munitions Company has perfected a new weapon, which consists of a long iron tube, out of which an iron ball containing explosives is expelled by exploding gunpowder. The projectile flies through the air and falls among the enemy, causing great panic and much damage.

A dispute has arisen between the Defense Department and the Wabash company regarding the range of the projectile. Investigate the range using the following simple model.

Let the mass of the projectile be W/g, where W is the weight in pounds and $g = 32.16$ ft/s/s is the acceleration of gravity. The drag force on the projectile is in the opposite direction to that of the velocity vector and is given by

$$F_D = \tfrac{1}{2}AC_D\rho V^2$$

where A is the cross-sectional area of the projectile, ρ is the air density, V is the velocity (ft/s), and C_D is the drag coefficient. (See Table 6.1.)

The dynamic equations for the projectile in flight are

$$m\frac{d^2y}{dt^2} = -F_D\sin\theta - mg$$

$$m\frac{d^2x}{dt^2} = -F_D\cos\theta$$

where x is measured horizontally and y vertically upward and θ is the angle measured from the x axis to the velocity vector, positive upward. The value of m is W/g. In solving the differential equations, it is not necessary to use $\sin\theta$ and $\cos\theta$ because F_D contains V^2. For instance, $\sin\theta$ is the ratio V_y/V. Therefore, it is easier to introduce the new variables V_x and V_y, which are the components of the velocity vector, and find the magnitude V using $V = \sqrt{V_x^2 + V_y^2}$.

It is best to write the program in stages by making the drag function return a zero value for F_D. Then you can check the range and the trajectory for vacuum conditions over a flat earth with the simple equations from beginning physics. After that you can add constant air density and, finally, variation of air density with altitude.

Maximum range occurs for 45° in the drag-free case, but that will change for the drag case. You will have to explore to find the launch angle for maximum range.

Use the data given from Table 6.1 for the drag coefficient C_D and from Project Problem 6.2 for the air density. (The sea-level air density can be taken as $\rho_0 = 0.002378$ slugs/ft^3.) Spline functions will give the most accurate results, but in the case of the air density you may wish to try fitting the data with an exponential function of the altitude.

Mach	C_D
0.0	0.50
0.4	0.52
0.8	0.66
1.2	0.93
1.6	1.03
2.0	1.01
2.4	0.99
2.8	0.97
3.2	0.95
3.6	0.93
4.0	0.92

Table 6.1, Ch. 6

To find the Mach number to use in finding C_D it is necessary to use the table given below. The Mach number is the ratio of the projectile speed to the speed of sound at the given altitude. The function can be approximated as piecewise linear with sufficient accuracy.

Altitude (ft)	Speed of sound in air (ft/sec)
0	1116
5,000	1097
10,000	1077
15,000	1057
20,000	1036
25,000	1015
30,000	995
35,000	973
40,000	968
. . .	. . . (Constant at 968)
65,000	968
70,000	971
75,000	974
80,000	977
85,000	981
90,000	984
95,000	988
100,000	991

Use the following data: Diameter of the projectile is 1.00 ft and the projectile weighs 150 lb. The muzzle velocity is $V = 2200$ ft/s.

PROJECT PROBLEM 8.3

A thin metal cooling fin is welded to the wall of a furnace. The drawing in Figure 8.6 shows the general configuration. The temperature in the fin can be described by

$$ks\frac{d^2T}{dx^2} = 2h(T - T_A)$$

where $T(x)$ is the fin temperature at a distance x from the wall, T_A is the air temperature, k is the thermal conductivity of the fin, s is the thickness of the fin, and h is a heat-transfer coefficient for convective cooling with free circulation of the air. The equation is derived by establishing a heat balance for a thin slice of the fin perpendicular to x. The heat flowing into the slice is equal to the heat flowing out to the next slice plus the heat lost to convection on the ends of the slice. (Temperature equilibrium is assumed. Also, the fin is assumed to be so long and thin that temperature variations are only important in the x direction.)

Solve the boundary-value problem for the fin temperature, where the wall temperature is T_W and where the end condition is that $-k\,dT/dx = h(T - T_A)$. This means that the heat loss from convection at the end of the fin is equal to the flow rate, where these rates are per unit area.

Let $T_W = 250°F$ and let $T_A = 80°F$. The widths of the fin are 6.0 in. and 12.0 in., and the thickness is 0.25 in. Two different materials, copper and steel, will be used, with conductivities of 221.0 and 25.7 Btu/h ft °F. Use both materials in both widths. Use $h = 1.10$ Btu/h ft^2 °F.

FIGURE 8.6

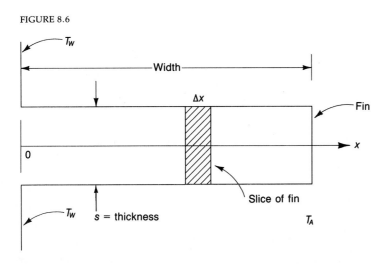

PROJECT PROBLEM 8.4

In damped simple harmonic motion, the damping force is assumed to be caused by viscous damping, in which the damping force is opposite in direction to the velocity vector and proportional in magnitude to the magnitude of the velocity. The differential equation is linear:

$$m\frac{d^2x}{dt^2} = -kx - b\frac{dx}{dt}$$

where m is the mass, k is the spring constant, and b is the coefficient of the viscous damping. The units might be slugs of mass, lbf/ft for the spring constant, and lbf s/ft for the damping coefficient. The equation can be solved by any of a number of classical methods.

Suppose that a mass attached to a spring slides on a surface so that the damping force is caused by Coulomb friction. The Coulomb friction force is constant and opposite in direction to the velocity vector. This also gives rise to a linear equation in the displacement, but the equation has a discontinuous function representing the friction and is tedious to solve in piecewise fashion.

The friction-force magnitude is given by the product of the normal force (if the surface is level, this is simply the weight) and the coefficient of kinetic friction μ_k. When the body is at rest, an applied force will cause a static frictional force to develop. This force cannot exceed the normal force times the coefficient of static friction μ_s, where $\mu_s > \mu_k$ in general. This means that the mass will come to rest and stay at rest at the first maximum point for which the spring force fails to overcome the static frictional force.

Solve this problem for a 1-lbm mass sliding on a surface with $\mu_k = 0.2$ and $\mu_s = 0.3$. The spring constant is $k = 2.0$ lbf/ft. The mass starts from rest with an initial displacement of 3.0 ft.

It may be advisable to devise a procedure to guard against integrating across the friction function discontinuity.

PROJECT PROBLEM 8.5

The circuit shown in Figure 8.7(a) is driven by a voltage-pulse generator whose output voltage v_1 is shown in Figure 8.7(b).

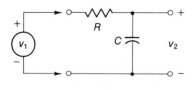

a.   b.

FIGURE 8.7

1. Show by the use of Kirchhoff's current law that the differential equation relating the output voltage v_2 to the input voltage v_1 is

$$\frac{v_2 - v_1}{R} + C\frac{dv_2}{dt} = 0$$

2. Solve the differential equation, assuming $v_2(0) = 0$, and show graphs of the output voltage v_2 for $t_2 = 5$, 10, and 15 ms. Show the solution for t from 0 to $2t_2$. Use $C = 1\ \mu F$ and $R = 1000\ \Omega$.

PROJECT PROBLEM 8.6

The circuit shown in Figure 8.8(a) is driven by a voltage-pulse generator whose output voltage v_1 is shown in Figure 8.8(b). D is a perfect diode that permits current to flow only from left to right through it, as shown in the figure.

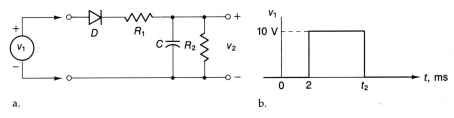

a.   b.

FIGURE 8.8

1. Show (using Kirchhoff's current law) that the differential equation relating the output voltage v_2 of the circuit to the input voltage v_1 of the circuit is

$$C\frac{dv_2}{dt} + \frac{v_2}{R_2} + \frac{v_2 - v_1}{R_1} = 0 \qquad \text{if } v_1 > v_2$$

$$C\frac{dv_2}{dt} + \frac{v_2}{R_2} = 0 \qquad \text{if } v_1 \leq v_2$$

2. Solve the differential equation numerically and show graphs of the output voltage v_2 for $t_2 = 5$, 10, and 15 ms. Show the solutions for t from $t = 0$ to $t = 2t_2$. Use $C = 1\ \mu F$ and $R_1 = R_2 = 1000\ \Omega$. Assume $v_2(0) = 0$.
3. Short out the diode so that current can flow in both directions and solve the problem numerically. Compare the solutions with those from (2).

PROJECT PROBLEM 8.7

Figure 8.9 shows a simple pendulum whose equation of motion is well known:

$$\frac{d^2\theta}{dt^2} = -\left(\frac{g}{L}\right)\sin\theta$$

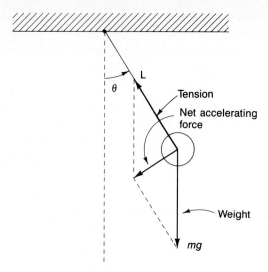

FIGURE 8.9   Simple pendulum.

where θ is the deflection of the pendulum from the vertical and where the pendulum swings in a vertical plane. For small angles the usual assumption is that $\sin \theta \simeq \theta$, which linearizes the problem so that it can be solved by classical methods. Although this procedure gives accurate results for small angles, it hides the fact that the period (the time for one complete swing) is amplitude dependent.

Solve the pendulum problem without linearization. Assume that $g = 9.8$ m/s^2 and that $L = 1$ m. Also, assume that the pendulum is released from rest at $\theta = \theta_0$ with $\dot{\theta}(0) = 0 = \dot{\theta}_0$.

Compare the solutions obtained from the nonlinear problem with those of the linearized problem for $\theta_0 = 5°$, $30°$, $60°$, and $90°$. One cycle of the motion with the longest period is sufficient for the comparison.

PROJECT PROBLEM 8.8

A sounding rocket of initial mass M is fired vertically through the earth's atmosphere. The applicable equation is

$$\frac{d(Mv)}{dt} = -Mg - \left(\frac{1}{2}\right) A\rho C_D v^2$$

where g is the acceleration of gravity, A is the cross-sectional area of the rocket, ρ is the air density, C_D is the drag coefficient, and v is the rocket velocity.

For a rocket in which the mass M is changing appreciably with the time, the time rate of change of the momentum Mv becomes

$$\frac{d(Mv)}{dt} = v'\frac{dM}{dt} + M\frac{dv}{dt}$$

where v' is the exhaust velocity with respect to the rocket. Thus, the differential equation becomes

$$M\frac{dv}{dt} = -Mg - \left(\frac{1}{2}\right)A\rho C_D v^2 - v'\frac{dM}{dt}$$

Solve this differential equation for a rocket of initial mass of 5000 lbm, in which the exhaust velocity of the burning fuel is 6000 ft/s (note that $v' < 0$ if velocities and distances are measured positive upward), and for which 4000 lbm of the rocket is fuel. The fuel is burned at a uniform rate and it is all used up in 30 s. The diameter of the rocket is 24 in. Use the drag coefficient whose values are given in Table 6.1 and the atmospheric density from Project Problem 6.2. (Assume that the sea-level air density is $\rho_0 = 0.002378$ slugs/ft^3.) Note that a function needs to be constructed to give M as a function of t. See Project Problem 8.2 for the variation of Mach number with altitude.

Listings

LISTING OF RKF45

```
{PROCEDURE RKF45 is a Runge-Kutta-Fehlberg combined fourth-
 and fifth-order ordinary differential equation solver for
 a system of NEQN first-order equations.

The call to RKF45 is

   RKF45(NEQN, YO, Y, TIN, TOUT, RELERR, ABSERR, FLAG)

where NEQN is the number of equations, TIN is the input
value of the independent variable, TOUT is the output
value of the independent variable, YO is a vector of
values of the dependent variables on input, Y is the
solution vector on output, RELERR and ABSERR are relative
and absolute desired error bounds on the solution, and
FLAG is a communication vector.

FLAG = 1 (input) requests integration from TIN to TOUT to
produce Y.  Normal termination sets FLAG = 2. FLAG should
be reset to 1 for the next call. FLAG = -1 (input) requests
integration over one RKF internal step toward TOUT.  Normal
termination sets FLAG = -2, and FLAG should be reset to -1
for another step.  The integrator will stop at TOUT
eventually.  FLAG = 3 on return indicates too many function
evaluations, FLAG = 4 indicates failure to pass the
accuracy test because the solution vanished and ABSERR was
zero, FLAG = 5 indicates too drastic an accuracy
requirement (increase the error tolerances), FLAG = 6
```

indicates too many output points, and FLAG = 7 indicates bad input data.

The user must provide a procedure DIFFEQ of the following form:

PROCEDURE DIFFEQ(T: REAL; VAR Y, YP: VECTOR);

. . .

BEGIN

  YP[1] := . . .
  YP[2] := . . .
  . . .

END;

The user must also:

1. Execute FINDU26 before calling RKF45 to compute a value for U26, 13 times the machine epsilon

2. Declare U26, ABSERR, RELERR, OLDABSERR, OLDRELERR, TIN, and TOUT as type REAL, and NEQN as a constant

3. Declare TYPE VECTOR = ARRAY[1..NEQN] OF REAL; and declare Y0, Y, and YP as VECTOR

4. Declare INIT as BOOLEAN, FLAG, OLDFLAG and EXITFLAG as INTEGER }

PROCEDURE FINDU26;

{Finds 13 times the machine epsilon}

VAR
   X: REAL;

BEGIN {FINDU26}

  X := 1.0;
  REPEAT
    X := X/2.0
  UNTIL (X + 1.0 = 1.0);
  U26 := 26.0*X

END;   {FINDU26}

```
PROCEDURE FEHL(      N:  INTEGER;
                   TIN:  REAL;
                     H:  REAL;
           VAR     Y0:  VECTOR;
           VAR    YP0:  VECTOR;
           VAR      T:  REAL;
           VAR      Y:  VECTOR;
           VAR     YE:  VECTOR  );

{Carries out an integration over step length H}

VAR
                    Y1:  VECTOR;
                    Y2:  VECTOR;
                    Y3:  VECTOR;
                    Y4:  VECTOR;
                    Y5:  VECTOR;
                   YP1:  VECTOR;
                   YP2:  VECTOR;
                   YP3:  VECTOR;
                   YP4:  VECTOR;
                   YP5:  VECTOR;
                     I:  INTEGER;
                    CH:  REAL;

BEGIN

  T  := TIN + H;
  CH := H/4.0;
  FOR I := 1 TO N DO Y1[I] := Y0[I] + CH*YP0[I];
  DIFFEQ(TIN + CH, Y1, YP1);

  CH := 3.0*H/32.0;
  FOR I := 1 TO N DO
    Y2[I] := Y0[I] + CH*(YP0[I] + 3.0*YP1[I]);
  DIFFEQ(TIN + 3.0*H/8.0, Y2, YP2);

  CH := H/2197.0;
  FOR I := 1 TO N DO
    Y3[I] := Y0[I]
          + CH*(1932.0*YP0[I] + (7296.0*YP2[I]
          - 7200.0*YP1[I]));
  DIFFEQ(TIN + (12.0/13.0)*H, Y3, YP3);

  CH := H/4104.0;
  FOR I := 1 TO N DO
    Y4[I] := Y0[I]
          + CH*((8341.0*YP0[I] - 845.0*YP3[I])
          + (29440.0*YP2[I] - 32832.0*YP1[I]));
  DIFFEQ(T, Y4, YP4);
```

```
            CH := H/20520.0;
            FOR I := 1 TO N DO
              Y5[I] := Y0[I]
                     + CH*((-6080.0*YP0[I]
                     + (9295.0*YP3[I] - 5643.0*YP4[I]))
                     + (41040.0*YP1[I] - 28352.0*YP2[I])));
            DIFFEQ(TIN + 0.5*H, Y5, YP5);

            CH := H/7618050.0;         {5th-order approximation}
            FOR I := 1 TO N DO
              Y[I] := Y0[I]
                    + CH*((902880.0*YP0[I]
                    + (3855735.0*YP3[I] - 1371249.0*YP4[I]))
                    + (3953664.0*YP2[I] + 277020.0*YP5[I])));

            CH := H/752400.0;
            FOR I := 1 TO N DO
              YE[I] := CH*((2090.0*YP0[I]
                     + (15048.0*YP4[I] - 21970.0*YP3[I]))
                     + (27360.0*YP5[I] - 22528.0*YP2[I])))

            {YE is the difference between the 5th- and 4th-
            order methods used for error control}

            END;     {FEHL}




            PROCEDURE  RKF45(    NEQN: INTEGER;
                             VAR   Y0: VECTOR;
                             VAR    Y: VECTOR;
                             VAR  TIN: REAL;
                             VAR TOUT: REAL;
                                RELERR: REAL;
                                ABSERR: REAL;
                             VAR FLAG: INTEGER  );

            {The main procedure}


            CONST     MAXNFE = 3000;

            VAR
                         H: REAL;
                        DT: REAL;
                      HMIN: REAL;
                   ABSFLAG: INTEGER;
                     PTCNT: INTEGER;
                       NFE: INTEGER;
```

```
        YP0:  VECTOR;
      ABORT:  BOOLEAN;
        OUT:  BOOLEAN;
      GOODH:  BOOLEAN;



FUNCTION NORM(VAR Y: VECTOR): REAL;

{Finds the infinity norm of vector Y}

VAR
          I:  INTEGER;
        MAX:  REAL;
       TEMP:  REAL;

BEGIN

  MAX := ABS(Y[1]);
  FOR I := 2 TO NEQN DO
    BEGIN
      TEMP := ABS(Y[I]);
      IF TEMP > MAX THEN MAX := TEMP
    END;
  NORM := MAX

END;     {NORM}



PROCEDURE CHECK_INPUT;

{Checks flags and input parameters}

BEGIN

  ABSFLAG := ABS(FLAG);
  IF (NEQN < 1)
     OR (ABSERR < 0)
     OR (RELERR < 0)
     OR (ABSFLAG = 0)
     OR (FLAG > 7)
     OR (TOUT = TIN) THEN
     BEGIN
       FLAG := 7;
       ABORT := TRUE
     END
  ELSE
    CASE FLAG OF
```

```
        -2, 2:  IF INIT THEN
                  CASE EXITFLAG OF
                    3:  NFE := 0;
                    4:  IF ABSERR = 0 THEN
                          ABORT := TRUE;
                    5:  IF (ABSERR <= OLDABSERR)
                          AND (RELERR <= OLDRELERR)
                            THEN
                              ABORT := TRUE;
                    0:     {No action}
                  END       {CASE EXITFLAG}
                ELSE FLAG := OLDFLAG;
          3:  BEGIN
                NFE := 0;
                FLAG := OLDFLAG
              END;
          4:  IF ABSERR > 0 THEN FLAG := OLDFLAG
              ELSE ABORT := TRUE;
      5, 6, 7:  ABORT := TRUE;
        -1, 1:  ABORT := FALSE
    END;    {CASE  FLAG}
    IF NOT ABORT THEN
      BEGIN
        OLDFLAG := FLAG;
        EXITFLAG := 0;
        OLDABSERR := ABSERR;
        OLDRELERR := RELERR;
        DT := TOUT - TIN
      END

END;    {CHECK_INPUT}



PROCEDURE START;

{START initializes the integrator}

VAR
          TEMP: REAL;
         TOLER: REAL;
             I: INTEGER;



PROCEDURE FIXSTEP;

{Estimates the starting step size}

VAR
     NORMYP0: REAL;
```

```
BEGIN     {FIXSTEP}

   INIT := TRUE;
   TOLER := RELERR*NORM(Y0) + ABSERR;
   H := ABS(DT);
   NORMYP0 := NORM(YP0);
   IF TOLER < NORMYP0*EXP(5*LN(H)) THEN
     BEGIN
       TEMP := ABS(TIN);
       IF TEMP > H THEN H := TEMP;
       H := U26*H;
       TEMP := EXP(0.2*LN(TOLER/NORMYP0));
       IF TEMP > H THEN H := TEMP
     END

END;     {FIXSTEP}


BEGIN     {START}

   IF ABSFLAG = 1 THEN
     BEGIN
       INIT := FALSE;
       PTCNT := 0;
       DIFFEQ(TIN, Y0, YP0);
       NFE := 1;
       IF TOUT = TIN THEN
         FLAG := 2
       ELSE FIXSTEP
     END;
   IF NOT INIT THEN FIXSTEP;
   IF DT < 0 THEN H := -H;
   IF ABS(H) > ABS(DT) THEN PTCNT := PTCNT + 1;
   IF PTCNT = 100 THEN FLAG := 6;
   IF ABS(DT) <= U26*ABS(TIN) THEN
     BEGIN
       FOR I := 1 TO NEQN DO Y0[I] := Y0[I] + DT*YP0[I];
       DIFFEQ(TOUT, Y0, YP0);
       NFE := NFE + 1;
       TIN := TOUT;            {Use Euler's method for a  }
       FLAG := 2              {short step                }
     END;
   OUT := FALSE

END;     {START}



PROCEDURE RKFS;

{Main integrator}
```

```
VAR
        I:  INTEGER;
        T:  REAL;
    TOLER:  REAL;
        YE:  VECTOR;


PROCEDURE STEPSIZE;

BEGIN     {STEPSIZE}

  GOODH := TRUE;
  HMIN := U26*ABS(TIN);      {Smallest allowable}
  DT := TOUT - TIN;                      {stepsize}
  IF ABS(DT) < 2*ABS(H) THEN
    IF 0.9*ABS(DT) > ABS(H) THEN
      H := 0.5*DT
    ELSE                     {Adjust stepsize to hit}
      BEGIN                  {output point.  Look    }
        OUT := TRUE;         {ahead two points        }
        H := DT
      END

END;     {STEPSIZE}



PROCEDURE INCREMENT;

{Manages step size changes}

VAR
      TEMP: REAL;
     ERROR: REAL;



PROCEDURE CARRY_ON;

{Carries out a step and checks step size}

BEGIN     {CARRY_ON}

  IF NFE > MAXNFE THEN
    BEGIN
      FLAG := 3;       {Too many function}
      EXITFLAG := 3;          {evaluations}
      ABORT := TRUE
    END
```

```
      ELSE
        BEGIN
          FEHL(NEQN, TIN, H, Y0, YP0, T, Y, YE);
          NFE := NFE + 5;
          TOLER := 0.5*(NORM(Y0) + NORM(Y))*RELERR + ABSERR;
          IF TOLER = 0 THEN
            BEGIN
              FLAG := 4;
              EXITFLAG := 4;      {Inappropriate error}
              ABORT := TRUE                {tolerance}
            END
      ELSE IF NOT ABORT THEN
        BEGIN
          ERROR := NORM(YE);
          IF ERROR >= TOLER THEN
            BEGIN
              GOODH := FALSE;    {Unsuccessful step;}
              OUT := FALSE;      {reduce step size  }
              IF ERROR >= 59049.0*TOLER THEN
                H := 0.1*H
              ELSE H := 0.9*H/EXP(0.2*LN(ERROR/TOLER));
              IF ABS(H) <= HMIN THEN
                BEGIN
                  FLAG := 5;       {If step size too small}
                  EXITFLAG := 5;
                  ABORT := TRUE
                END
            END      {IF ERROR}
        END      {IF NOT ABORT}
      END      {ELSE}

END;          {CARRY_ON}


BEGIN    {INCREMENT}

  REPEAT
    CARRY_ON
  UNTIL ABORT OR (ERROR < TOLER);
  IF NOT ABORT THEN
    BEGIN
      TIN := T;
      Y0 := Y;
      DIFFEQ(TIN, Y0, YP0);
      NFE := NFE + 1;
      IF GOODH THEN            {Choose next step size}
        BEGIN
          IF ERROR <= (1.889568E-4)*TOLER THEN
            TEMP := 5          {Increment step size}
```

```
                    ELSE TEMP := EXP(0.2*LN(0.9*TOLER/ERROR));
                    TEMP := TEMP*ABS(H);
                    IF HMIN > TEMP THEN TEMP := HMIN;
                    IF H >= 0 THEN H := TEMP ELSE H := -TEMP
              END     {IF GOODH}
        END     {IF NOT ABORT}     {Step size is not       }
                                   {incremented if step-   }
    END;     {INCREMENT}           {size failure occurred  }
                                   {on previous step       }


  BEGIN     {RKFS}

    REPEAT
      STEPSIZE;
      INCREMENT
    UNTIL ABORT OR OUT OR (FLAG <= 0);
    IF OUT THEN
      BEGIN
        TIN := TOUT;
        FLAG := 2
      END
    ELSE IF NOT ABORT THEN FLAG := -2

  END;     {RKFS}




  BEGIN     {RKF45}

    CHECK_INPUT;
    IF NOT ABORT THEN
      BEGIN
        START;
        RKFS
      END

  END;     {RKF45}
```

9

Mathematical Functions

9.1 Introduction

In this chapter we will look at problems associated with the computation of mathematical functions. We often take for granted the standard functions built into PASCAL or FORTRAN because they seem to work so well without any special consideration on our part. The construction of good approximations to mathematical functions for use in the computer is a real art. We will look at some of the methods used and the important considerations that must be observed in order to create good computer functions.

The first consideration is efficiency. We would like the computation to take as little time as possible. If we remember that there may be hundreds or thousands of computers, each executing millions of sine functions, for instance, each day, then it is apparent that it is worthwhile to spend a reasonable amount of time in creating efficient function subprograms. Efficiency also requires economy of memory space, which was extremely important in early computers with very limited memory. The idea of storing tables of function values and doing interpolation (which might have been faster) was unthinkable. Hence computing a function value on demand became the usual method, which we still use today.

The second important consideration is accuracy. We would like the computer functions to give values that have acceptable errors. What we would like, of course, is to have no more error than the machine ϵ over the entire domain of the argument of the function. Such accuracy is usually achieved in the case of the standard functions in most good compiler systems, although it is a good idea to check out the standard functions on the system you are using. (Some manufacturers will provide the accuracy information in their publications.) In many cases it may not be necessary, or even practical, to have full machine accuracy for the function values, but it is important to know the error bounds. In this chapter we will look at several methods of approximating functions, including methods we saw in Chapter 6, and then develop methods that permit us to place bounds on the errors of these approximations.

9.1.1 Standard Functions

Standard functions are those functions absolutely necessary for mathematical computation. These are usually considered to be the square root, natural logarithm, exponential, sine, cosine, and inverse tangent functions. Some software systems include other functions such as the tangent, inverse sine and cosine, logarithm to the base 10, and hyperbolic functions. Values of all these functions can be obtained from the standard functions. For example, we can write $\log_{10}(x) = \ln(x)/\ln(10)$ and $\sinh(x) = (e^x - e^{-x})/2$. The user may find it convenient to create sets of such functions when the need arises.

Fortunately for the creator of the standard functions, it is often necessary to develop approximations over only a limited portion of the domain. For example, e^x can be evaluated by an approximation over [0, 1] for all values of x. The limited domain needed results from the rules of exponents from algebra, which allow us to write $e^{-x} = 1/e^x$ and, for instance, $e^{2.563} = e^2 e^{0.563}$. Since every exponent can be written as the sum of an integer and a fraction, we need only a good approximation of e^x over [0, 1] and a good value of e to raise to integer powers.

Sines and cosines can be evaluated using an accurate approximation over [0, $\pi/4$]. For x outside this range, we can use a combination of reduction of the angle by multiples of 2π and a manipulation of trigonometric identities.

Likewise, for the natural logarithm, we can write the argument x as $x = e^n q$, where q is a fraction such that $e^{-1} < q \leq 1$ and n is an integer. Then $\ln(x) = \ln(e^n q) = n + \ln(q)$. Thus we need only a good approximation for the logarithm on the domain $[e^{-1}, 1]$.

Finally, for the square root, we realize that in floating point, the argument always has a fraction f, where $\frac{1}{2} \leq f < 1$, and an exponent that is an even or odd integer. If the exponent e is even, then the square root is simply $(f)^{1/2}$ times $2^{e/2}$. However, if e is odd, then we can make it even by letting $e = e + 1$ and letting $f = f/2$. We can now take the square root the same way except that the new f can be as small as $\frac{1}{4}$. Thus the approximation necessary for the square root needs to be accurate over $[\frac{1}{4}, 1]$.

Abramowitz and Stegun [1964] contains examples of many mathematical functions including all of the standard ones. In all cases the error bounds and the domains over which the approximations apply are given.

9.1.2 Polynomial Approximations

The kinds of approximations we will study are polynomial approximations. Rational functions, which are quotients of polynomials, are also important but are somewhat beyond the scope of this book. They are particularly important in repre-

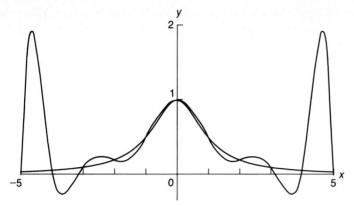

FIGURE 9.1 Interpolating polynomial approximation of
$$y = \frac{1}{(1 + x^2)}.$$

senting functions that have horizontal or vertical asymptotes. Polynomials are useful because they involve only addition, subtraction, and multiplication, which are among the basic arithmetic operations the computer can perform. Furthermore, polynomials have mathematical properties that we will find useful and they can be evaluated efficiently if they are in nested (Horner's rule) form.

We might try using interpolating polynomials, such as we saw in Chapter 6, but we will find that there is not an easy way to estimate the error of such an approximation, let alone control it. Figure 9.1 shows again the effect of using an interpolating polynomial through the points $x = -5, -4, \ldots, 4, 5$ on the function $f(x) = 1/(1 + x^2)$. It gives very poor results, particularly near the ends of the interval. We would like to be able to "tame" this polynomial, and we can do so to the extent shown in Figure 9.2. This approximation was produced by

FIGURE 9.2 Remes algorithm approximation of $y = \dfrac{1}{(1 + x^2)}$
with tenth-degree polynomial.

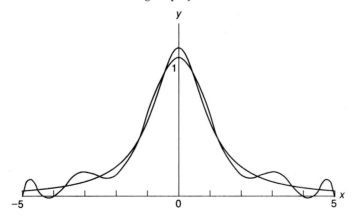

the use of the Remes algorithm, which we will study later in the chapter. Here we have given up the fixed interpolating points in favor of reducing the maximum error over the interval. We note that the largest error is greatly reduced at the expense of increasing the errors in the middle of the region. A better polynomial approximation than this requires using a higher-degree polynomial (or going to a rational function approximation, which will fit perfectly in this case!).

Because of the lack of control over the error, we will not use interpolating polynomials for function approximation if we wish to impose error bounds on the approximations. Instead, we will go on to look at methods that try to control the maximum error.

9.2   Least-Squares Approximation of Functions

We shall look at the least-squares approximation method because it is one way of attempting to limit the error in creating a function approximation. The method is easy to understand from a simple example.

Example 9.1

Approximate e^x over $[0, 1]$ using $p_2(x) = ax^2 + bx + c$ in the sense of least squares.

To do this, we construct the integral analog of the sum of the squared deviations in the least-squares formulation we saw in Chapter 6.

$$F(a, b, c) = \int_0^1 (ax^2 + bx + c - e^x)^2 \, dx$$

To minimize $F(a, b, c)$ with respect to a, b, and c, we need to find the first partial derivatives of F with respect to a, b, and c, set the derivatives equal to zero, and solve the system of linear simultaneous equations. To find the derivatives we must apply Leibniz's rule, the conditions for the existence of which are satisfied in this case, and differentiate under the integral sign:

$$\frac{\partial F}{\partial a} = 2 \int_0^1 (ax^2 + bx + c) x^2 \, dx = 0$$

$$\frac{\partial F}{\partial b} = 2 \int_0^1 (ax^2 + bx + c) x \, dx = 0$$

$$\frac{\partial F}{\partial c} = 2 \int_0^1 (ax^2 + bx + c) 1 \, dx = 0$$

which gives, in a different order,

$$a \int_0^1 x^2 \, dx + b \int_0^1 x \, dx + c \int_0^1 1 \, dx = \int_0^1 e^x \, dx$$

$$a \int_0^1 x^3 \, dx + b \int_0^1 x^2 \, dx + c \int_0^1 x \, dx = \int_0^1 x e^x \, dx$$

$$a \int_0^1 x^4 \, dx + b \int_0^1 x^3 \, dx + c \int_0^1 x^2 \, dx = \int_0^1 x^2 e^x \, dx$$

After evaluation of the integrals we have

$$\begin{bmatrix} \frac{1}{3} & \frac{1}{2} & 1 \\ \frac{1}{4} & \frac{1}{3} & \frac{1}{2} \\ \frac{1}{5} & \frac{1}{4} & \frac{1}{3} \end{bmatrix} \begin{bmatrix} a \\ b \\ c \end{bmatrix} = \begin{bmatrix} e - 1 \\ 1 \\ e - 2 \end{bmatrix}$$

The solution yields the polynomial

$$p_2(x) = 0.83918\,39828 x^2 + 0.85112\,50463 x + 1.01299\,1311$$

The error curve, $e^x - p(x)$, is shown in Figure 9.3. Note that the error is not uniformly bounded ("equal ripple") but is worse at the ends of the interval. □

Least-squares function approximations suffer at least two disadvantages. One is that they are not uniformly bounded in the error. The second is that there is no easy way to find the error bound without enumerating the error point by point across the interval. For these reasons we will do no more with the least-squares method of approximation of functions. We will move on to look at two methods that give more satisfactory results.

FIGURE 9.3 Error of least-squares fit of $y = ax^2 + bx + c$ to e^x over interval $[0, 1]$.

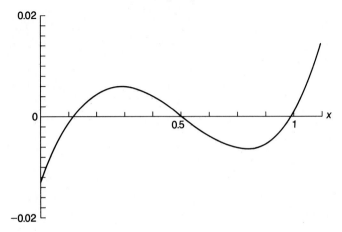

9.3   Chebyshev Economization

In Chapter 7 we looked briefly at the Chebyshev polynomials $T_i(x)$, where we saw that one definition of $T_i(x)$ was

$$T_i(x) = \cos(i \operatorname{Cos}^{-1} x)$$

From this definition we see that each $T_i(x)$ is bounded:

$$|T_i(x)| \leqslant 1, \quad -1 \leqslant x \leqslant 1$$

This is a valuable property of the Chebyshev polynomials because it allows us to make function approximations that are almost minimax. By a *minimax* approximation we mean the following:

Let $f(x)$ be a function to be approximated over $[a, b]$, and let $p(x)$ be a polynomial of degree n. The maximum error of $p(x)$ in approximating $f(x)$ is

$$\max_x |f(x) - p(x)|$$

The minimax approximation is the one that minimizes this maximum error. A theorem by Weierstrass guarantees the existence of an approximating polynomial of this kind:

If $f(x)$ is continuous on the interval $[x_1, x_2]$, then given an $\epsilon > 0$, there exists an n which depends on ϵ and a polynomial of degree n, $p_n(x)$, such that

$$|f(x) - p_n(x)| < \epsilon$$

for all x in $[x_1, x_2]$.

We make such an approximation by adjusting the coefficients of the polynomial until the error curve (difference between the function and the polynomial) has maxima and minima that are the same magnitude (*equal ripple*). The virtue of a minimax approximation is that it has the same maximum error everywhere in $[a, b]$. In other words, we can depend upon the sine function to have no more error than, for instance, 10^{-12}, on the interval $-\pi/4$ to $+\pi/4$. This cannot be said of some other kinds of approximations, such as least squares or interpolation.

Although Chebyshev economizations do not give true minimax approximations, for many purposes these approximations are good enough. Furthermore, the economization procedure allows us to set the error bound and then create a polynomial approximation that satisfies the bound. The construction of such an approximation can best be explained through an example.

Example 9.2

Suppose we would like to approximate e^x on the interval $[-1, 1]$ by means of a polynomial with guaranteed error over the interval of no more than 10^{-3}. Furthermore, we would like to use the minimum number of polynomial terms necessary.

We start by using Taylor's formula with remainder, with the point of expansion being $x = 0$. (This is also known as Maclaurin's series with remainder because the point of expansion is zero.)

$$e^x = 1 + x + \frac{x^2}{2} + \frac{x^3}{6} + \frac{x^4}{24} + \cdots + \frac{x^n}{n!} + \left(\frac{x^{n+1}}{(n+1)!}\right) f^{(n+1)}(z)$$

where $x < z < 0$ or $0 < z < x$. We cannot find the exact value of z, but we can find an upper bound to $f^{(n+1)}(z)$ and thus an upper bound to the error term. $f^{(n+1)}(z) = e^z$, and this has its maximum magnitude at $z = 1$. Hence the maximum absolute value of the remainder is

$$\left|\frac{x^{n+1}e}{(n+1)!}\right| = \frac{e}{(n+1)!}$$

Now let us require that $e/(n+1)!$ be considerably smaller than the error bound 10^{-3}, say less than 10^{-4}. In that case

$$\frac{e}{(n+1)!} < 10^{-4}$$

or

$$(n+1)! > 10^4 e = 27182.81828. \ldots$$

Now $6! = 720$, $7! = 5040$, and $8! = 40,320$, so that $n = 7$ will satisfy the inequality. This means that

$$p_7(x) = 1 + x + \frac{x^2}{2} + \cdots + \frac{x^7}{7!}$$

has less than 10^{-4} error over the interval $[-1, 1]$ (actually the error is less than 6.742×10^{-5}). However, the error is concentrated at the ends of the interval and the accuracy is better than necessary in the middle. Figure 9.4 shows the error of the truncated series in approximating the function.

We can write x^7 in terms of the Chebyshev polynomial

$$T_7(x) = 64x^7 - 112x^5 + 56x^3 - 7x$$

as

$$x^7 = \left(\frac{1}{64}\right)[T_7(x) + 112x^5 - 56x^3 + 7x]$$

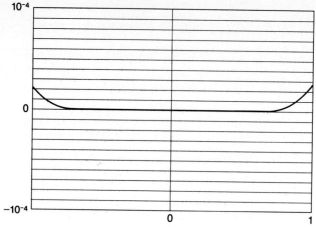

FIGURE 9.4   Error of $p_7(x)$.

Then

$$p_7(x) = 1 + x + \cdots + \frac{x^6}{6!} + \left(\frac{1}{64}\right)\frac{T_7(x) + 112\,x^5 - 56\,x^3 + 7x}{5040}$$

$$= 1 + \frac{46{,}081}{46{,}080}x + \cdots + \frac{x^6}{6!} + \frac{T_7(x)}{322{,}560}$$

Now the $T_7(x)$ term has a coefficient of 3.1×10^{-6}, which is well below 10^{-4} in value. Because $|T_7(x)| \le 1$ over the interval $[-1, 1]$, we can drop this term with very little degradation of accuracy anywhere in the interval. When we drop the $T_7(x)$ term, we are left with

$$p_6(x) = 1 + \left(\frac{46{,}081}{46{,}080}\right)x + \left(\frac{1}{2}\right)x^2 + \left(\frac{959}{5760}\right)x^3$$

$$+ \left(\frac{1}{24}\right)x^4 + \left(\frac{25}{2880}\right)x^5 + \left(\frac{1}{720}\right)x^6$$

Figure 9.5 shows the error of approximation using $p_6(x)$. The error curve shows only a small change from that of $p_7(x)$.

We now repeat the process with x^6. From

$$T_6(x) = 32x^6 - 48x^4 + 18x^2 - 1$$

we find that

$$x^6 = \frac{T_6(x) + 48x^4 - 18x^2 + 1}{32}$$

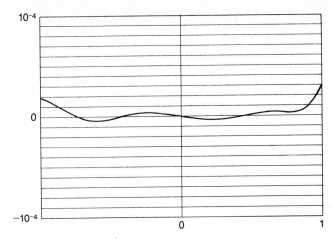

FIGURE 9.5   Error of $p_6(x)$.

so that

$$p_6(x) = 1 + \frac{46{,}081}{46{,}080}x + \frac{1}{2}x^2 + \frac{959}{5760}x^3 + \frac{1}{24}x^4 + \frac{25}{2880}x^5$$
$$+ \left(\frac{1}{32}\right)\frac{T_6(x) + 48x^4 - 18x^2 + 1}{720}$$

The coefficient of $T_6(x)$ is 4.3403×10^{-5} so that we can drop the term and have no more total error than

$$6.742 \times 10^{-5} + 3.100 \times 10^{-6} + 4.3403 \times 10^{-5} = 1.139 \times 10^{-4}$$

Thus, if we drop the $T_6(x)$ term, we will still have only about 1.14×10^{-4} error with a fifth-degree polynomial:

$$p_5(x) = \frac{23{,}041}{23{,}040} + \left(\frac{46{,}081}{46{,}080}\right)x + \left(\frac{639}{1280}\right)x^2$$
$$+ \left(\frac{959}{5760}\right)x^3 + \left(\frac{21}{480}\right)x^4 + \left(\frac{25}{2880}\right)x^5$$

Figure 9.6 shows the error of approximation of $p_5(x)$, which shows that the accuracy of the approximation is actually better than the indicated bound by a factor of almost two. The actual error has increased substantially, but we have two fewer terms in our approximation.

Because $T_5(x) = 16x^5 - 20x^3 + 5x$, we can now write

$$p_5(x) = \frac{23{,}041}{23{,}040} + \cdots + \left(\frac{25}{16}\right)\frac{T_5(x) + 20x^3 - 5x}{2880}$$

The coefficient of $T_5(x)$ is 5.43×10^{-4}, which when added to the total error to

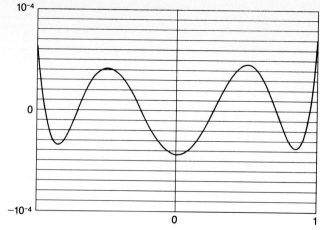

FIGURE 9.6   Error of $p_5(x)$.

date gives about 6.57×10^{-4} total error if we drop $T_5(x)$. When we drop $T_5(x)$, we have left

$$p_4(x) = \frac{23{,}041}{23{,}040} + \left(\frac{11{,}489}{11{,}520}\right)x + \left(\frac{639}{1280}\right)x^2 + \left(\frac{2043}{11{,}520}\right)x^3 + \left(\frac{1008}{23{,}040}\right)x^4$$

Figure 9.7 shows that the error has now increased by a factor of about 10 as a result of dropping the fifth-power term.

FIGURE 9.7   Error of $p_4(x)$.

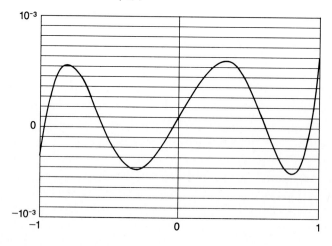

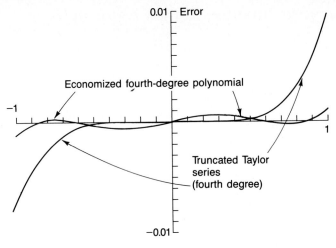

FIGURE 9.8

On the next use of this procedure, we find that the error added by dropping the $T_4(x)$ term is greater than we can accept and still keep the error below 10^{-3}. Therefore, we cannot reduce the polynomial any more. The final polynomial is

$$p(x) = 1.00004\,3403 + 0.99752\,60417x + 0.49921\,8750x^2$$
$$+ 0.17647\,56944x^3 + 0.04375x^4$$

Figure 9.8 shows the error between e^x and $p_4(x)$ as a function of x over $[-1, 1]$. Also shown is the graph of error for the truncated Taylor series

$$y(x) = 1 + x + \frac{x^2}{2!} + \frac{x^3}{3!} + \frac{x^4}{4!}$$

Note that this approximation is in error by more than 10^{-3} when $|x| > 0.392$. At $x = 1$ the error is about 9.95×10^{-3}, which is about ten times the allowable error, and is about 17 times the error of the economized polynomial at that point. □

The question arises, What if we wish to approximate a function over some interval other than $[-1, 1]$? The problem becomes a little more difficult for a general interval, but the principle is still the same. Perhaps the easiest way is to transform the interval desired, say $[a, b]$, into $[-1, 1]$ by means of the linear transformation $x = (b - a)z/2 + (b + a)/2$, where z is the variable on the interval $[-1, 1]$. After the polynomial in z has been economized, the variable z is replaced by the old variable x by means of $z = [2x - (b + a)]/(b - a)$. This may require a good deal of tedious algebra.

9.4   Minimax Approximation: Remes Algorithms

In the previous section we saw how to economize a polynomial in order to produce a near-minimax polynomial approximation. In this section we shall look at a method that iteratively approaches a minimax solution whether the approximating function is a polynomial or is a rational function. In the rational case it involves the solution of a mildly nonlinear system, whereas in the case of a polynomial it requires the solution only of a linear system.

Minimax approximations oscillate around the function being approximated. In general, the endpoints of the interval are among the points where the maximum absolute errors occur. The number of such maxima is equal to the degree of the approximating polynomial plus 2, the proof of which is beyond the level of this book.

First we choose a set of $n + 2$ points on $[-1, 1]$ using $x = -1$ and $x = 1$ and n arbitrary points in between. It is reasonable to take equally spaced points to start, but it may be better to use the Chebyshev points given by $\cos[i\pi/(n + 1)]$, $i = 0, \ldots, n + 1$.

We then solve the set of equations

$$f(x_i) - p(x_i) = (-1)^i E, \qquad i = 1, 2, 3, \ldots, n + 2$$

where $f(x)$ is the function being approximated, $p(x)$ is the approximating polynomial, and E is the error at each of the selected points x_i. We force the errors to be of alternating sign and equal magnitude. Once we have found values for the polynomial coefficients and E, we can form the polynomial $p(x)$ and evaluate it at any point on $[-1, 1]$. We will find that there are points for which $|p(x) - f(x)|$ is greater than E because the x_i were chosen arbitrarily. We now find the value z for which $p(x)$ has the maximum absolute value of the error and substitute it for the previously used adjacent point at which the error has the same sign. If $z < x_0$ (which could happen if the endpoint $x = -1$ gets replaced in the course of the computation) then z replaces x_0 if the errors at z and at x_0 are of the same sign. However, if the errors are of opposite sign, then z is added to the set x_i, and the last point at the opposite end of the interval is dropped. The same kind of procedure is followed if $z > x_{n+1}$, which could result in x_0 being dropped if z is to be added to the set instead of replacing a nearby point.

Once z has been exchanged for an old point (hence the name sometimes used, the *Remes exchange algorithm*), the linear system is solved again and a new point z is found; the cycle is repeated until no point can be found at which the error exceeds E in magnitude. A practical stopping criterion is to quit when the maximum error magnitude exceeds E by less than 1%, for instance. Such a polynomial is so close to being optimal that it is usually not worth the extra effort to continue the process.

It is not necessary that the interval be $[-1, 1]$. Any closed interval will do. The Chebyshev points on $[-1, 1]$ can be mapped onto any interval $[a, b]$ for use as starting points by means of a linear transformation if it is desired to use them instead of linearly spaced points.

Example 9.3

Find a quadratic polynomial $p(x) = ax^2 + bx + c$ that is a minimax approximation to e^x on $[-1, 1]$.

We will start by choosing only a very small discrete set of points x:

$$x = -1.00,\ -0.75,\ -0.50,\ -0.25,\ 0.00,\ 0.25,\ 0.50,\ 0.75,\ 1.00$$

in order to simplify the problem. (The computer can only represent a finite set of numbers between -1 and 1, anyway, except that the possible set is vastly larger than this set!) We will arbitrarily choose -1.00, -0.25, 0.25, and 1.00 as our initial $n + 2$ points. The equations to be solved are

$$a -\quad b + c + E = 0.36787\,94412$$
$$0.0625a - 0.25b + c - E = 0.77880\,07831$$
$$0.0625a + 0.25b + c + E = 1.28402\,5417$$
$$a +\quad b + c - E = 2.71828\,1828$$

The solution of this system is

$$a = 0.54577\,87035$$
$$b = 1.14225\,0808$$
$$c = 0.99730\,19311$$
$$E = -0.03295\,038512$$

Using these values of a, b, and c, we can find the errors at the points:

-1.00	$+0.03295$
-0.75	-0.02475
-0.50	-0.04391
-0.25	-0.03295
0.00	-0.00270
0.25	$+0.03295$
0.50	$+0.05615$
0.75	$+0.04399$
1.00	-0.03295

The largest error occurs at 0.50, so we replace 0.25 in our set of trial points and solve the system

$$a -\quad b + c + E = 0.36787\,94412$$
$$0.0625a - 0.25b + c - E = 0.77880\,07831$$
$$0.2500a + 0.50b + c + E = 2.11700\,0017$$
$$a +\quad b + c - E = 2.71828\,1828$$

for

$$a = 0.55815\,222676 \qquad |E| = 0.04223$$

$$b = 1.13297\,06663$$

$$c = 0.98492\,840842$$

The new set of errors is

-1.00	$+0.04223$
-0.75	-0.02321
-0.50	-0.04855
-0.25	-0.04223
0.00	-0.01507
0.25	$+0.01903$
0.50	$+0.04223$
0.75	$+0.03162$
1.00	-0.04223

The largest error (and the only one larger than the ones at the solution points) is at -0.50, so we substitute -0.50 for -0.25 because the errors at those points have the same sign. We then solve

$$a - \quad b + c + E = 0.36787\,94412$$

$$0.25a - 0.50b + c - E = 0.60653\,06597$$

$$0.25a + 0.50b + c + E = 1.64872\,1271$$

$$a + \quad b + c - E = 2.71828\,1828$$

to find

$$a = 0.55393\,95590 \qquad |E| = 0.04434$$

$$b = 1.13086\,4333$$

$$c = 0.98914\,10756$$

and a set of errors

-1.00	$+0.04434$
-0.75	-0.01978
-0.50	-0.04434
-0.25	-0.03775
0.00	-0.01086
0.25	$+0.02245$
0.50	$+0.04434$
0.75	$+0.03188$
1.00	-0.04434

Now there are no errors greater in magnitude than the errors at the solution points. No improvement can be made in the polynomial fit with this number of points. It might have been possible to have obtained a better fit had we used a thousand points or more in the set of points, but that would have made the example unwieldy. Figure 9.9 is a plot of the error across the interval, showing that the approximation is quite good and that it appears to be equal ripple or very close to it.

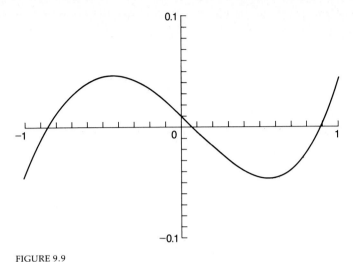

FIGURE 9.9

Note that the error E increases during the process as we introduce points at which the error is greater in magnitude than E at the previous step.   □

The Remes algorithm can also be used to provide minimax approximations for rational functions, say $p(x)/q(x)$, where $p(x)$ and $q(x)$ are polynomials. In this case it can be shown that there will be at least $n + 2$ error extrema in the approximation interval, where n is the sum of the degree of $p(x)$ and the degree of $q(x)$. The method is the same as before. For a set of $n + 2$ values of x, solve to find the coefficients in the polynomials $p(x)$ and $q(x)$ and the value of E from

$$f(x_i) - \frac{p(x_i)}{q(x_i)} = (-1)^i E$$

Unfortunately, this system is nonlinear, but fortunately it can be solved iteratively. To do this we assume a value for E_0 and solve for the coefficients and E_1 in the linear system

$$p(x_i) - [f(x_i) - (-1)^i E_0] q(x_i) = f(x_i) - (-1)^i E_1$$

Solution of this set is repeated to convergence. Then the set of errors is examined as before to find a point at which the error is a maximum. That point is then exchanged for one of the $n + 2$ points, using the same criteria as in the polynomial case. This is an example of an iteration loop contained within another iteration loop. Such processes are often slow, but this one has not proved to have great convergence problems.

A final word is required concerning using the Remes algorithms for finding approximations to functions where the approximations must be accurate to the full machine word length. In such cases it is necessary to use double-precision arithmetic in the algorithm. To produce double-precision approximations, it is necessary to extend the precision or to accept some inaccuracy in the values. Providing accurate values of the function to be approximated often requires using many terms from the infinite series expansion.

9.5   Procedure MINIMAX

The procedure MINIMAX is a PASCAL Remes procedure to approximate a function by a polynomial over some domain [a, b] in minimax fashion. It is called from a driver program, which needs to provide to the procedure some parameters and variable definitions.

The call to MINIMAX is as follows:

    MINIMAX (M, N, XL, XR, A, E, MAX)

where

M	=	the number of points to be used in the interval of interpolation
N	=	the degree of the interpolating polynomial
XL, XR	=	the left and right abscissas of the interval
A	=	the array of coefficients of the polynomial on return
E	=	the last value of E found in the procedure
MAX	=	the last calculated maximum error at any of the M points; MAX < 1.01E

The calling program must establish the following data types:

```
TYPE DATAVECTOR = ARRAY[0..M] OF REAL;
TYPE COEFVECTOR = ARRAY[0..N] OF REAL;
TYPE INTVECTOR  = ARRAY[0..N+1] OF INTEGER;
```

The DATAVECTOR is used for the storage of the points in the interval, storage of the values of the function at those points, storage of the values of the interim polynomial at the points, and storage of the errors at the points. COEFVECTOR is used to store the coefficients of the approximating polynomial. The array PIUSE of type INTVECTOR stores the indices of the points from the larger set (size M + 1) used in the algorithm. These "points in use" are the ones that are successively replaced in the Remes algorithm until the minimax approximation is achieved.

It is necessary to provide "include" statements in the calling program to read in the source code for GAUSS and SOLVE, which are used to solve the linear equations. (SOLVE should always precede GAUSS because SOLVE is called from within GAUSS in finding the condition number.) MINIMAX prints a warning message if the condition number exceeds 1,000,000.

The choice of the value of M is not critical, but it must be greater than N + 2. Good results can be obtained with values on the order of a few hundred. The following are the coefficients and the value of MAX obtained for approximating e^x on [0, 1] with a third-degree polynomial and M = 100, 200, and 800:

	100	200	800
A[0]	9.99456 31169 E-01	9.99455 93072 E-01	9.99456 00251 E-01
A[1]	1.01660 18912 E+00	1.01659 31018 E+00	1.01659 42112 E+00
A[2]	4.21682 66357 E-01	4.21709 03167 E-01	4.21705 70347 E-01
A[3]	2.79997 27370 E-01	2.79979 69497 E-01	2.79981 91377 E-01
MAX	5.46339 60099 E-04	5.46676 63790 E-04	5.43997 48835 E-04

There are only very slight changes in the coefficients from one approximation to the other and no evident changes in the graphs of the error.

9.6   IMSL Subroutine IRATCU

FORTRAN subroutine IRATCU produces a weighted, rational minimax approximation of an arbitrary function. The rational approximation may itself depend upon a continuous, monotonic function PHI (X), which is supplied by the user.

The call to IRATCU is

        CALL IRATCU (F, PHI, G, A, B, L, M, P, Q, WK, IER)

F is the user-supplied function to be approximated. It must be a function of a single variable and must be declared EXTERNAL in the calling program. PHI is a user-supplied function, also declared EXTERNAL. For a plain rational approximation the user defines PHI (X) = X in an EXTERNALly declared function. G is a weight function, which can be used to give more emphasis to some part of the region than to others. It must be continuous, nonvanishing on [A, B], and must also be declared EXTERNAL. An equally weighted approximation should use G (X) = 1.0. If G is chosen as |F (X)|, then the approximation will minimize the relative error. (G must never be zero, however.) A and B are the bounds on the interval, with A < B.

The rational function is specified in degree by L and M, where L is the degree of the numerator polynomial and M is the degree of the denominator polynomial. For a plain polynomial, approximation M is set to 0. P and Q are arrays in which the polynomial coefficients are returned. P returns the numerator coefficients with the constant term in P (1). Q returns the denominator coefficients with the constant term in Q (1). P must have dimension at least L + 1 and Q must have dimension at least M + 1. WK is a work array of length at least (L + M + 8) * (L + M + 2). On return, WK (1) contains a number whose magnitude is the minimax error in the approximation.

IER is the error parameter, and has the following return values:

IER =   0   Normal termination.

IER = 129   Failed to converge. Often indicates a common factor between numerator and denominator. Try reducing L and M by one.

IER = 130   Failed to converge. Possibility of a pole (zero of the denominator) in [A, B]. Try reducing L and M or changing [A, B].

IER = 131   Failed to converge in 20 iterations.

IER = 132   Failed to converge—linear system was singular. Perhaps degenerate approximation. Try changing L and M.

IER =   33   (Warning.) Error was reduced as far as possible. May be a good approximation but not minimax.

Exercises

1. Use the Chebyshev economization method to find an approximation to $\sin x$ on $[-1, 1]$ that has no more than 10^{-3} error on the interval.

2. Use the Chebyshev economization method to find an approximation to $\ln(1 + x) = x - \frac{1}{2}x^2 + \frac{1}{3}x^3 + \cdots$, which converges on $-1 < x < 1$. The error should not exceed 10^{-3} over the interval $-\frac{1}{2} < x < \frac{1}{2}$.

3. Draw graphs of the errors on $[-\pi/2, \pi/2]$ for the Taylor series approximations (about $x = 0$) of $\cos x$ when the series are truncated at one, two, three, and four terms. Then truncate $\cos x$ to five terms and produce the Chebyshev economized polynomials of one, two, three, and four terms. Compare the graphs.

4. Calculate a minimax approximation to e^x on $[-1, 1]$ using the rational function $(ax + b)/(x + c)$. Note that one coefficient in a rational function can always be made equal to 1.

5. Calculate a polynomial approximation to $1/x$ over $[1, 2]$ with an error no greater than 0.01 in magnitude. This may require several runs using MINIMAX to find a polynomial of sufficiently high degree.

6. Use MINIMAX to find cubic polynomial approximations to the following functions:
   (a) $\ln x$, $1/e < x < 1$
   (b) $\sin x$, $-\pi/4 < x < \pi/4$
   (c) $\cos x$, $0 < x < \pi$
   (d) $\tan x$, $-\pi/4 < x < \pi/4$
   (e) $\tan^{-1}x$, $-1 < x < 1$

   Plot the error curves for each of the approximations.

7. The following table gives the density of water at various temperatures. Suppose for a computer program you need a very simple function to describe the density of water versus temperature (a cubic polynomial, at the most). What maximum error will you have if you approximate the relationship with polynomials of first, second, or third degree?

Hint: Use MINIMAX with 569 points, one for each degree Fahrenheit from 32° to 600°. Set up the array of function values so that the missing points are linear interpolations between the given points from the table. Then create minimax polynomials of degrees 1, 2, and 3 for this set of data. Note that MAX in each case is at least as great as the error at any data point.

Temperature, °F	Density, lbm/ft³
32	62.4
40	62.43
50	62.4
60	62.3
70	62.3
80	62.2
90	62.1
100	62.0
150	61.2
212	60.0
250	58.8
300	57.3
350	55.6
400	53.6
450	51.6
500	49.0
550	45.9
600	42.4

8. Create PASCAL function subprograms to add to the standard functions the following functions:

   (a) $\sinh(x)$, $\cosh(x)$, and $\sinh^{-1}(x)$
   (b) $\log_{10}(x)$ and $\log_2(x)$
   (c) B^x, where $1 < B$
   (d) $\sin^{-1}(x)$ and $\cos^{-1}(x)$
   (e) $\tan_4^{-1}(x, y)$, where $\tan_4^{-1}$ identifies the quadrant in which the point (x, y) lies and gives its angle correctly in the range $[-\pi, \pi]$. Use the standard arctangent function as the basis, but arrange the function so that it will work correctly for all (x, y). This will require testing for $x = 0$ and handling that case as a special one.

9. Show how the sine and cosine of any angle x can be computed using an accurate approximation for the sine function over the interval $[0, \pi/4]$. The value of x can be positive or negative and can be of any size.

Problems

PROJECT PROBLEM 9.1

Starting from MINIMAX, create a program to find minimax rational approximations. Chapter 7 of Ralston and Rabinowitz [1964] is a good reference. This treatment uses the second Remes algorithm, which provides for replacement of all of the solution points by points at which there is greater error instead of replacing only one point at a time. You may wish to incorporate that method instead of the first Remes algorithm.

Check out your program for $f(x) = e^{-x}$ on [0, 2].

PROJECT PROBLEM 9.2

Find a minimax approximation to the air density table in Project Problem 6.2. Try $n = 3, 4,$ and 5. Give the estimated maximum error in each case.

PROJECT PROBLEM 9.3

Revise MINIMAX in such a way that the procedure can be sent an array of points representing the function to be approximated instead of having the array generated internally in MINIMAX by a function defined in the calling program.

Listings

LISTING OF MINIMAX

```
PROCEDURE MINIMAX (M:        INTEGER;
                   N:        INTEGER;
                   XL:       REAL;
                   XR:       REAL;
         VAR       A:        COEFVECTOR;
         VAR       E:        REAL;
         VAR       MAX:      REAL        );
```

```
{MINIMAX attempts to find the minimax error polynomial
 approximation to a user-supplied function.   The method
 used is the first Remes algorithm.   The program terminates
 when the computed maximum MAX is within 1% of the value of
 ERR.

 The calling program must contain a type declaration

         TYPE DATAVECTOR    =    ARRAY[0..M] OF REAL;

 where M + 1 is the number of points to be used over the
 interval.   M should normally be a large multiple of N, the
 degree of the approximating polynomial.

 A similar declaration is necessary for the array of
 polynomial coefficients
```

```
      TYPE COEFVECTOR   =   ARRAY[0..N] OF REAL;
```

The user must also define a real function (of a single real variable) called FUN:

```
      FUNCTION FUN(X: REAL): REAL;
```

The parameters in the call are

M   =   Number - 1 of points to be used;
         $M \gg N + 2$

N   =   Degree of approximating polynomial

XL, XR = Left and right bounds of the interval over
           which the approximation is to be performed

A   =   Return COEFVECTOR of polynomial coefficients
         (A[0] contains the constant term)

E   =   Last value of the error calculated by the
         Remes algorithm

MAX =   Maximum of $|P(x) - FUN(x)|$ at one of the M
         data points. $MAX < 1.01*|ERR|$

The calling program MUST contain 'include' statements that allow GAUSS and SOLVE to be loaded with MINIMAX. The proper order in which these must be included is:

1.   SOLVE, which must precede GAUSS for
     compilation

2.   GAUSS

3.   MINIMAX   }

VAR
```
            I,J:      INTEGER;
          ERRS:       DATAVECTOR;
         PIUSE:       INTVECTOR;
             F:       DATAVECTOR;
             P:       DATAVECTOR;
          DONE:       BOOLEAN;
          IMAX:       INTEGER;
             S:       INTEGER;
          COND:       REAL;
```

```
          PROCEDURE FILL_FUN_ARRAY;


          {Procedure FILL_FUN_ARRAY produces the set of M + 1
           ordinates F[I] which represent the function FUN being
           approximated.}


          VAR
                         I:    INTEGER;
                         X:    REAL;

          BEGIN

            FOR I := 0 TO M DO
              BEGIN
                X := XL + I*(XR - XL)/M;
                F[I] := FUN(X);
              END;
          END;



          FUNCTION POLY ( X:    REAL ): REAL;


          {Function POLY evaluates the current approximating
           polynomial at any abscissa in the interval of
           approximation.}


          VAR
                         I:    INTEGER;
                         P:    REAL;

          BEGIN

            P := A[N];
            FOR I := N-1 DOWNTO 0 DO
              P := P*X + A[I];
            POLY := P;

          END;



          PROCEDURE EVAL_POLY;


          {Procedure EVAL_POLY evaluates the current version of the
           polynomial at the M + 1 abscissas and stores the values in
           P[I].}
```

```
VAR
                    I:      INTEGER;
                    X:      REAL;
              LENGTH:      REAL;


BEGIN

  LENGTH := XR - XL;
  FOR I := 0 TO M DO
    BEGIN
      X := XL + I*LENGTH/M;
      P[I] := POLY(X);
    END;
END;




PROCEDURE FIND_MAX( VAR MAX:       REAL;
                    VAR IMAX:      INTEGER);


{Procedure FIND_MAX finds the maximum (MAX) of
|F[I] - P[I]| for the current version of the
polynomial and also finds the index (IMAX) of
the abscissa for that point.  (The F[I] - P[I]
are stored in ERRS[I].)}


VAR
                    I:      INTEGER;
                 DIFF:     REAL;

BEGIN

  IMAX := 0;
  MAX := 0.0;
  FOR I := 0 TO M DO
    BEGIN
      DIFF := ABS(ERRS[I]);
      IF MAX < DIFF THEN
        BEGIN
          IMAX := I;
          MAX  := DIFF;
        END;
    END;
END;



PROCEDURE   COMPUTE_ERRORS;
```

```
{Procedure COMPUTE_ERRORS produces
 ERRS[I] := P[I] - F[I] for I := 0 to M.}


VAR
                     I:     INTEGER;


BEGIN

  FOR I := 0 TO M DO
    ERRS[I] := P[I] - F[I];

END;




PROCEDURE LOAD_MATRIX;


{Procedure LOAD_MATRIX fills the matrix and vector to be
 sent to GAUSS and SOLVE to find the A[I] (polynomial
 coefficients) and the value of E.}


VAR              I:     INTEGER;
                 J:     INTEGER;
                 T:     REAL;
                 X:     REAL;
                 E:     REAL;


BEGIN

  FOR I := 0 TO N+1 DO
    BEGIN
      T := 1.0;
      X := XL + PIUSE[I]*(XR - XL)/M;
      FOR J := 0 TO N DO
        BEGIN
          C[I+1,J+1] := T;
          T := T*X;
        END
    END;

  IF ODD(N) THEN E := 1.0 ELSE E := -1.0;
  FOR I := 0 TO N+1 DO
    BEGIN
      C[I+1,N+2] := E;
      E := -E;
    END;
```

```
   FOR I := 0 TO N+1 DO
      B[I+1] := F[PIUSE[I]];

END;



PROCEDURE EXCHANGE;


{Procedure EXCHANGE finds the appropriate member of PIUSE
 (points in current use) to be replaced by the point with
 index IMAX and then exchanges them.}



VAR
                    I:    INTEGER;
                    J:    INTEGER;

BEGIN

   I := -1;
   REPEAT
     I := I + 1
   UNTIL
     (IMAX < PIUSE[I]) OR (I > N+1);

   IF I = 0 THEN
     BEGIN
       IF ERRS[IMAX]*ERRS[PIUSE[I]] < 0 THEN
         BEGIN
           FOR J := N+1 DOWNTO 1 DO
             PIUSE[J] := PIUSE[J-1];
           PIUSE[I] := IMAX
         END
       ELSE PIUSE[I] := IMAX
     END
   ELSE IF I > N+1 THEN
     BEGIN
       IF ERRS[IMAX]*ERRS[PIUSE[N+1]] < 0 THEN
         BEGIN
           FOR J := 1 TO N+1 DO
             PIUSE[J-1] := PIUSE[J];
           PIUSE[N+1] := IMAX
         END
       ELSE PIUSE[N+1] := IMAX
     END
```

```
      ELSE
        BEGIN
          IF ERRS[IMAX]*ERRS[PIUSE[I]] > 0 THEN
            PIUSE[I] := IMAX
            ELSE PIUSE[I-1] := IMAX
        END

END;




PROCEDURE INIT_PIUSE;


{Procedure INIT_PIUSE selects equally spaced abscissas for
 the initialization of the PIUSE vector.}



VAR
                   I:    INTEGER;

BEGIN

  PIUSE[0]    := 0;
  PIUSE[N+1] := M;
  FOR I := 1 TO N DO
    PIUSE[I] := I*(M DIV (N+1))

END;


PROCEDURE TRANSFER;


{Procedure TRANSFER removes the contents of the B vector
 returned as the solution from GAUSS/SOLVE and stores the
 values in the polynomial coefficient vector A.   B[N + 2]
 is the value of E.}



VAR
                   I:    INTEGER;

BEGIN

  FOR I := 0 TO N+1 DO
    A[I] := B[I+1];

END;
```

```
BEGIN {Procedure MINMAX}

  FILL_FUN_ARRAY;
  INIT_PIUSE;
  S := N + 2;
  DONE := FALSE;

  REPEAT
    LOAD_MATRIX;
    GAUSS(S,C,COND,IPVT);
    IF COND > 1.0E+06 THEN WRITELN('Condition = ',COND);
    SOLVE(S,C,B,IPVT);
    TRANSFER;
    EVAL_POLY;
    COMPUTE_ERRORS;
    FIND_MAX(MAX,IMAX);
    IF MAX < 1.01*ABS(B[N+2]) THEN DONE := TRUE;
    EXCHANGE;
  UNTIL DONE;

  E := ERRS[IMAX];


END;
```

Answers to Selected Exercises

CHAPTER 1

1. **(a)** $35.75_{10} = 43.6_8$; **(c)** $39.3_{10} = 47.23146\ldots_8$; **(f)** $1{,}000{,}000_{10} = 3{,}641{,}100_8$
2. **(a)** $32.75_{10} = 20.C_{16}$; **(c)** $43.3_{10} = 2B.4CC\ldots_{16}$; **(f)** $1{,}000{,}000_{10} = F4{,}240_{16}$
3. **(a)** $525.37_8 = 341.0484375_{10}$; **(c)** $1001.05_8 = 513.078125_{10}$; **(f)** $1{,}000{,}000_8 = 262{,}144_{10}$
4. **(a)** $A3F.2B_{16} = 2623.16796875_{10}$; **(c)** $100.0C_{16} = 256.046875_{10}$; **(f)** $1{,}000{,}000_{16} = 16{,}777{,}216_{10}$

Comment on Problems 5 and 6: For most of us it is easier to convert from the first base to decimal and then from decimal to the second base. This is not necessary if you can do arithmetic in one of the bases other than decimal.

5. **(a)** $37.37_8 = 1F.7C_{16}$; **(c)** $1001.05_8 = 201.14_{16}$; **(f)** $1{,}000{,}000_8 = 40{,}000_{16}$
6. **(a)** $A3F.2B_{16} = 5077.125_8$; **(c)** $1001.5_{16} = 10{,}001.014_8$; **(f)** $1{,}000{,}000_{16} = 100{,}000{,}000_8$

Note: Decimal points mark the breaks between exponents and fractions.

8. **(a)** 55.1393; **(c)** -47.3966; **(d)** 51.5000
9. **(a)** $-53.1254 \rightarrow -125.4$; **(e)** $50.9600 \rightarrow 0.96$
10. **(a)** $53.3651 \rightarrow 365.1$; **(d)** $-57.3384 \rightarrow -3.384 \times 10^6$
11. **(a)** $52.6572 \rightarrow 65.72$; **(e)** $-51.1018 \rightarrow -1.018$
12. 8 bits, with a bias of 10000000_2
18. **(a)** 56.2218 (first word); 669029 (second word)

CHAPTER 2

1. **(a)** 315.66; **(e)** 6754, or 6.754×10^3
2. $37.345 \rightarrow 37.3$ (correct); $37.345 \rightarrow 37.35 \rightarrow 37.4$ (incorrect)

3. **(a)** Calculator result: 2.226094855 in.^{3}
   Round correctly to: 2.23 in.^{3}
   Interval arithmetic: (2.205546418, 2.246770527)

4. **(a)** 1.29; **(c)** 15.48; **(f)** 47.34

5. 0.1554171883 rad $\simeq$ 8.905°

6. **(a)** |Error| = 0.0782913822 (truncated series); |error| = 0.3084251375 (Taylor formula estimate); **(d)** |Error| = 0.005170918 (truncated series); |error| = 0.00555258546

CHAPTER 4

3. The inverse is: $\begin{bmatrix} -100 & 100 \\ 101 & -100 \end{bmatrix}$

5. Condition number: 745.9
   Solution: 9.0000000007 E+00
              -3.6000000004 E+01
                 3.0000000004 E+01

6. Condition number: 1.8217 E+04
   Solution: 1.6000000013 E+01
              -1.2000000013 E+02
                 2.4000000030 E+02
              -1.4000000019 E+02

7. Condition number: 9.4037 E+05
   Solution: 6.2999998249 E+02
              -1.2599999648 E+04
                 5.6699998429 E+04
              -8.8199997577 E+04
                 4.4099998798 E+04

9. Solution: 5, 2, -1, 2

10. Solution: -1.5, 0.5, 5, -5.5, -1.5

CHAPTER 5

11. $a = 0.3600349828$

15. $x = 0.8526055020$

17. **(a)** Roots: 3, -2, -9; **(b)** roots: $\pm 3i$, 7, -5

18. **(c)** First 3 roots: 0, 4.493409458; 7.725251837; **(d)** root: 0.3499696317

21. Radius = 1.567769041 ft

23. Solution for continuous variable N: 9.506396114 (Quicksort is faster for $N = 10$ but slower for $N = 9$.)

24. **(a)** roots: $-1 \pm 2i$, $-2 \pm i$; **(d)** roots: 0.961142, -1.244423, 1.955072, -2.562164 $\pm 5.385950i$; **(f)** roots: $-0.809017 \pm 0.587785i$, 0.309017 $\pm 0.951057i$

25. Each dimension should be increased by 2.095503788 in.

CHAPTER 6

8. The entry for $x = 0.80$ should be 0.88811 or 0.88812.

9. **(a)** Cos 40°40′ $= 0.75851$ (Use $\sin^2 x + \cos^2 x = 1$.)

10. If t is used as 45, 46, . . . , then the speed s is predicted as $s = 37.72749 + 1.692806t$. The predicted speed for 1985 is 181.616 mi/h.

12. $n(0.63) = 1.457156$
    $n(0.53) = 1.460860$
    $n(0.43) = 1.467231$

13. **(a)** $q = 342.9103211/(1 + 0.19633722t)$

14. Constant in front of the exponential is 3.476086242.

16. **(a)** Standard error of estimate versus degree of polynomial:

Degree	Standard error
1	1.533396
2	0.6951708
3	0.2345778 ←
4	0.2305954
5	0.2817315

It does not appear that a polynomial higher than third degree is necessary.

CHAPTER 7

8. $I = 6.271376001$

10. **(a)** $I = 1.605412977$; **(c)** $I = 0.7963427209$

13. Selected values:

X	Gamma (X)
1.0	1.00000 00000
1.5	0.88622 69255
2.0	1.00000 00000

15. **(a)** $T = 4829.3619$ K

16. Area $= 3565.3692$ in.^{2}

17. Main lobe: 2.836303; first side lobes: 0.1480193

CHAPTER 8

3. General solution is
$$y = \frac{1}{1 - Ce^{x^2/2}}$$

For $y(0) = 1 + t$, t small, C becomes $C = t/(1 + t)$. Thus for $t = 0$, the solution reduces to $y(x) = 1$.

6. $y(1) = 2.5723411$, $y(2) = 4.1693662$

7. $y(1) = 0.7615942$, $y(2) = 0.9640276$

8. $y(1) = 1.25$, $y(2) = 1.0$, $y(5) = 3.25$

9. $y(1) = -0.5$, $y(2) = -1.0$

10. $y(1) = 0.33333333$, $y(2) = 0.50000000$

This equation has a singularity at the origin. Replace the right-hand side of the equation in the neighborhood of the origin by F^4/x^4, where $F = y(0) + y'(0)x + y''(0)x^2/2$. Use this in place of $(y/x)^4$ for $0 \leqslant x < 0.001$.

CHAPTER 9

1. $\sin x \cong (1915/1920)\, x - (300/1920)\, x^3$ max $|\text{error}| \cong 0.0006$

6. **(a)** $-2.3458339024 + 1.2872765846x - 3.8423014551x^2$ 500 points; $|\text{Max}| = 1.87424 \times 10^{-3}$ **(c)** $0.99552108632 + 6.9822365401 \times 10^{-2}x - 0.67188626229x^2$ 500 points; $|\text{Max}| = 4.53151 \times 10^{-3}$

References

The addresses of the organizations from which substantial numerical software libraries discussed in Chapter 3 can be obtained are given, along with the addresses of publishers of references to these libraries.

SOFTWARE LIBRARIES

IMSL, Inc.
7500 Bellaire Boulevard
NBC Building, Floor 6
Houston, Texas 77036

National Energy Software Center
Argonne National Laboratory
Argonne, Illinois 60439

REFERENCE MATERIAL

LINPACK User's Guide, J. J. Dongarra, C. B. Moler, J. R. Bunch, and G. W. Stewart, SIAM Publications, Society for Industrial and Applied Mathematics, 117 South 17th Street, Philadelphia, Pennsylvania 19130.

"Matrix Eigensystems Routines—EISPACK Guide," B. T. Smith et al., *Lecture Notes in Computer Science No. 6,* 2d ed., Springer-Verlag New York, Inc., 175 Fifth Avenue, New York, New York 10010.

"The FUNPACK Package of Special Function Subroutines," William J. Cody, *ACM Transactions on Mathematical Software* 1 (1975): 13–25.

"A Practical Guide to Splines," Carl de Boor, *Lecture Notes in Applied Mathematics No. 27*, Springer-Verlag New York, Inc., 175 Fifth Avenue, New York, New York 10010, 1979. (PPPACK)

Numerical Recipes, William H. Press, Brian P. Flannery, Saul A. Teukolsky, and William T. Vetterling, Cambridge University Press, 32 East 57th Street, New York, New York 10022, 1986. This book is not directly related to any of the program libraries just listed, but it is an extensive reference for a wide variety of numerical procedures. Both FORTRAN and PASCAL program listings are printed in the book, and floppy disks are available for either the FORTRAN or the PASCAL version of the programs.

Bibliography

In addition to the references cited in the text, the bibliography includes several more advanced books on numerical analysis that are frequently cited in the literature as standard references.

Abramowitz, M., and I. A. Stegun. 1964. *Handbook of Mathematical Functions With Formulas, Graphs, and Mathematical Tables.* National Bureau of Standards Applied Mathematics Series No. 55, Washington, D.C.: U.S. Government Printing Office.

Acton, F. S. 1970. *Numerical Methods that Work.* New York: Harper and Row.

Atkinson, K. E. 1978. *An Introduction to Numerical Analysis.* New York: John Wiley.

Brent, R. 1973. *Algorithms for Minimization Without Derivatives.* Englewood Cliffs, N.J.: Prentice-Hall.

Cavanagh, J. J. F. 1984. *Digital Computer Arithmetic: Design and Implementation.* New York: McGraw-Hill.

Churchill, R. V., and J. W. Brown. 1984. *Complex Variables and Applications.* 4th ed. New York: McGraw-Hill.

Cline, A. K., C. B. Moler, G. W. Stewart, and J. H. Wilkinson. 1979. An Estimate for the Condition Number of a Matrix. *SIAM J. on Num. Anal.* 16(2): 368–375.

Coddington, E. A. 1961. *An Introduction to Ordinary Differential Equations.* Englewood Cliffs, N.J.: Prentice-Hall.

Collier, R. S., E. A. Monash, and P. F. Hultquist. 1981. Modeling Natural Gas Reservoirs—A Simple Model. *Soc. of Petr. Engrs. J.* 21(5): 521–526.

Dahlquist, G., and A. Bjorck. 1974. *Numerical Methods.* Trans. Ned Anderson. Englewood Cliffs, N.J.: Prentice-Hall.

de Boor, C. 1971. "CADRE: An Algorithm for Numerical Quadrature." In *Mathematical Software*, ed. J. R. Rice. New York: Academic Press.

Dongarra, J. J., C. B. Moler, J. R. Bunch, and G. W. Stewart. 1979. *LINPACK User's Guide.* Philadelphia: SIAM Publications.

Fehlberg, E. 1970. Klassische Runge-Kutta-Formeln Vierter und Niedrigerer Ordnung mit Schrittweiten-Kontrolle und ihre Anwendung auf Warmeleitungsprobleme. *Computing* 6: 61–71.

Forsythe, G. E. 1957. Generation and Use of Orthogonal Polynomials for Data-fitting on a Digital Computer. *J. SIAM* 5: 74–88.

Froberg, C.-E. 1985. *Numerical Mathematics.* Menlo Park, Calif.: Benjamin/Cummings.

Gabel, R. A., and R. A. Roberts. 1980. *Signals and Linear Systems.* 2d ed. New York: John Wiley.

Gear, C. W. 1971. *Numerical Initial Value Problems in Ordinary Differential Equations.* Englewood Cliffs, N.J.: Prentice-Hall.

Guttman, I., and S. S. Wilks. 1965. *Introductory Engineering Statistics.* New York: John Wiley.

Hamming, R. W. 1959. Stable Predictor-Corrector Methods for Ordinary Differential Equations. *J. Assoc. Computing Mach.* 6: 37–47.

Hastings, C., Jr. 1955. *Approximations for Digital Computers.* Princeton, N.J.: Princeton University Press.

Hildebrand, F. B. 1974. *Introduction to Numerical Analysis.* New York: McGraw-Hill.

Householder, A. S. 1953. *Principles of Numerical Analysis.* New York: McGraw-Hill.

Hultquist, P. F., E. A. Monash, and K. Norman. 1978. "Simulating a Gas-Water Reservoir, Part 1: Mathematical Model and Phase-Plane Solutions." In *Simulation Councils Proceedings Series,* vol. 8, no. 2, ed. K. E. F. Watt. La Jolla, Calif.: Society for Computer Simulation.

Hultquist, P. F., and E. A. Monash. 1979. "A Mathematical Model of a Gas-Water Reservoir." In *Information Linkage between Applied Mathematics and Industry,* ed. P. C. C. Wang. New York: Academic Press.

Isaacson, E., and H. B. Keller. 1966. *Analysis of Numerical Methods.* New York: John Wiley.

Jensen, K., and N. Wirth. 1974. *PASCAL User Manual and Report.* 2d ed. New York: Springer-Verlag.

Johnson, L. W., and R. D. Riess. 1982. *Numerical Analysis.* 2d ed. Reading, Mass.: Addison-Wesley.

Kernighan, B. W., and P. J. Plauger. 1978. *The Elements of Programming Style.* 2d ed. New York: McGraw-Hill.

Knuth, D. E. 1973. *The Art of Computer Programming.* Vol. 3, *Sorting and Searching.* Reading, Mass.: Addison-Wesley.

Ledgard, H. 1986. *Professional Pascal.* Reading, Mass.: Addison-Wesley.

Nilsson, J. W. 1985. *Electric Circuits.* 2d ed. Reading, Mass.: Addison-Wesley.

Press, W. H., B. P. Flannery, S. A. Teukolsky, and W. T. Vetterling. 1986. *Numerical Recipes.* New York: Cambridge University Press.

Ralston, A., and P. Rabinowitz. 1978. *A First Course in Numerical Analysis.* 2d ed. New York: McGraw-Hill.

Rice, J. R. 1964, 1969. *The Approximation of Functions.* Vol. 1, *Linear Theory.* Vol. 2, *Advanced Topics.* Reading, Mass.: Addison-Wesley.

Savitch, W. J. 1986. *An Introduction to the Art and Science of Programming: TURBO Pascal Edition.* Menlo Park, Calif.: Benjamin/Cummings.

Scott, N. R. 1985. *Computer Number Systems and Arithmetic.* Englewood Cliffs, N.J.: Prentice-Hall.

Shampine, L. F., and M. K. Gordon. 1975. *Computer Solution of Ordinary Differential Equations*. San Francisco: W. H. Freeman.

Simmons, G. F. 1972. *Differential Equations with Applications and Historical Notes*. New York: McGraw-Hill.

Slider, H. C. 1976. *Practical Petroleum Reservoir Engineering Methods*. Tulsa, Okla.: The Petroleum Publishing Company.

Traub, J. F. 1964. *Iterative Methods for the Solution of Equations*. Englewood Cliffs, N.J.: Prentice-Hall.

Wilkinson, J. H. 1959. The Evaluation of the Zeros of Ill-conditioned Polynomials, Part 1. *Numerische Mathematik* 1: 150–166.

Wilkinson, J. H. 1963. *Rounding Errors in Algebraic Processes*. Englewood Cliffs, N.J.: Prentice-Hall.

Young, D. M. 1950. Iterative Methods for Solving Partial Difference Equations of the Elliptic Type. Ph.D. diss., Harvard University.

Young, D. M., and R. T. Gregory. 1972. *A Survey of Numerical Mathematics*. Vols. 1 and 2. Reading, Mass.: Addison-Wesley.

Index